QUANTUM LEAPS

QUANTUM

LEAPS

:: Jeremy Bernstein ::

THE BELKNAP PRESS OF
HARVARD UNIVERSITY PRESS
Cambridge, Massachusetts, and London, England
2009

Printed in the United States of America

Chapter 1 was originally published, in somewhat different form,
as "Auden's Cheery Bishop," *The New Criterion*, 27 (October
2008).

Library of Congress Cataloging-in-Publication Data

Bernstein, Jeremy, 1929–
Quantum leaps / Jeremy Bernstein.
p. cm.
Includes bibliographical references and index.
ISBN 978-0-674-03541-6 (alk. paper)
1. Quantum theory—Miscellanea.   I. Title

QC174.13.B47 2009
530.12—dc22   2009013778

Acknowledgments

THIS IS SUCH an eclectic book and so many people have been kind enough to make suggestions and criticisms that I would like to distribute these acknowledgments by topic. On Auden I would like to thank Freeman Dyson, Nicholas Jenkins, Arthur Kirsch, Edward Mendelson, Oliver Sacks, and Elizabeth Sifton. On the Dalai Lama I am grateful to Walter Isaacson and Abner Shimony. On all things Bohmian I am grateful to Ken Ford, Murray Gell-Mann, Basil Hiley, and David Pines. On Léon Rosenfeld I am grateful to Loren Graham and Anja Skaar Jacobsen. On Newtonian matters I am grateful to Alan Shapiro and on Philipp Frank to Gerald Holton. On the vexed subject of the quantum theory of measurements I have had lively interactions with Elihu Abrahams, Steve Adler, Andy Cohen, Steve Gasiorowicz, Jim Hartle, Eugen Merzbacher, and Bill Unruh. I have also had helpful comments from Peter Kaus, Arthur Miller, Bruce Rosenblum, and Oliver Sacks. At Harvard University Press, my thanks go to Michael Fisher, who kept the faith, and to Maria Ascher, for her encouragement and thoughtful editing. I would also like to thank my colleagues at the Aspen Center of Physics for many helpful discussions.

I wish that I could acknowledge my debt to the late John Bell in person. His spirit hovers over this book.

Contents

Whatever properties may ultimately be assigned to the atom, there is one which cannot be omitted—its power to seize and captivate the human mind. In fact, if we judged by the output of the printing press in the last few years, we might not unfairly assume that no sooner does anyone fall within the sphere of influence of this radiating personality, than he is seized with an irresistible determination to go home and write a book about it. Nor is the proselytising zeal confined to the pure physicist, whose protégé the atom may be presumed to be. We have books on the atom, some of them well done by chemists, by mathematicians, by technicians, and by journalists, and addressed to all sorts and conditions of readers. Thus we have *Atoms for Amateurs, Atoms for Adepts, Atoms for Adolescents, Atoms for Archdeacons, All about Atoms for Anybody*—these are not exact titles, but they indicate the scope of the volumes well enough—in fact, there seems to be a determination that no class of reader shall be left without an exposition of the subject suited to his condition and attainments. As these volumes continue to pour forth—we must assume that there are purchasers and readers. If we add to these the enormous output of serious scientific contributions from the many laboratories engaged in investigating the structures and properties of the atom, it is clear that this infinitesimal particle exerts an attraction unique in the history of science over the minds and imaginations of many types of men [and women].

—"The Atom Again," *Nature*, 118 (1926): 365

An Introduction

> For those who participated, it was a time of creation; there was terror as well as exaltation in their new insight. It will probably not be recorded very completely in history. As history, its recreation would call for an art as high as the story of Oedipus or the story of Cromwell, yet in a realm of action so remote from our common experience that it is unlikely to be known to any poet or any historian.
>
> —J. Robert Oppenheimer, *Science and the Common Understanding*, 1953 BBC Reith Lectures

WHEN OPPENHEIMER MADE these Delphic comments, I was just beginning a serious study of the quantum theory to which he refers. Not that I had any broad knowledge, but I would have agreed with him that at that time the quantum theory and its history were "unlikely to be known to any poet or historian." Things have changed. The last time I looked, there were more than 9.3 million entries on Google under the rubric "quantum theory," and these certainly included poets and historians, as well as film critics and Buddhist monks. It would take a serious cultural historian, which I am not, to trace what happened in those fifty-odd years. Having lived through this transformation, while I have some theories, I am not sure that I can pinpoint exactly when it occurred. I am, however, quite sure that it occurred not because more people were learning the quantum theory, but because more people were learning *about* the quantum theory and especially about its interpretation and foundations. I am also quite sure that at the time Oppenheimer gave his lectures, only a tiny number of physicists had any interest in the foundations. I will give an illustrative anecdote.

I spent two years at the Institute for Advanced Study in Princeton, beginning in the fall of 1957. Oppenheimer was the director. We had a weekly seminar, which he attended. By the spring, most of the obvious speakers had spoken and it was getting more difficult to recruit new ones. A young and gifted colleague of mine was recruited on the basis of a pre-print he had written on the quantum theory of measurement, to this day a highly controversial subject. Physicists disagree as to how measurements, which are described by classical physics, fit into the quantum theory. It was all but unheard of for a young physicist to work on something like this. It was not even considered physics. I admired his courage. I do not think that he asked to give a seminar. I think he was told. Oppenheimer was notorious for cutting down seminar speakers who he thought were wasting his time. At this seminar he outdid himself. The speaker had gotten out about five sentences when Oppenheimer said that Niels Bohr had answered all those questions in the 1930s and that the speaker had nothing to add. The seminar came to an end right there. I am glad to report that my colleague published his paper and that he went on to have a distinguished career.

I contrast this with another scene I witnessed thirty-odd years later, in the 1980s. It took place at Columbia University. John Bell, who I think is largely responsible for the renewed interest in this subject, had been invited to give a colloquium. It was meant to address an audience of the entire Physics Department, and it was held in the largest lecture hall available. The place was packed—standing room only. People came from every university in New York City and its surroundings—

even some from New Jersey. I can assure you no one stopped Bell to tell him that Bohr had done it all.

The purpose of this book is to give an account of that cultural transformation. The subject matter is eclectic, ranging from the Dalai Lama to W. H. Auden. I try to explain the relevant parts of the theory as I go along. There is essentially no mathematics. Words cannot really replace mathematics —but on a subject like this, mathematics cannot really replace words. There is a good deal of autobiography here. I hope the reader will not find this intrusive, but that is the kind of writer I am. I am sure that I have not covered the entire subject. There is too much for any one person to cover. But I hope that I have covered enough to make it clear that this subject is now known to poets and historians, as well as playwrights, novelists, filmmakers, Buddhist monks, and Communist ideologues, not to mention physicists.

:: **1** ::

Bishops

Rhyme-royal's difficult enough to play.
But if no classics as in Chaucer's day,
At least my modern pieces shall be cheery
Like English bishops on the Quantum Theory.

—W. H. Auden, "Letter to Lord Byron," 1936

Ah, my Lord of Birmingham, come in, sit on the
fire and anticipate the judgement of the Universal
Church.

—The Right Reverend Herbert Hensley[1]

I like the company of men of science: they are not
excessively intellectual in their hours of leisure
and they give good dinners.

—Bishop of Birmingham E. W. Barnes[2]

IN THE FALL OF 1957, I began what turned out
to be a two-year membership at the Institute for
Advanced Study in Princeton. This was a banner
time for physics. In 1956, the Chinese-born American physicists T. D. Lee and C. N. Yang noted in

several papers that what is called "parity symmetry"—the symmetry between left- and right-
handed descriptions—a cornerstone of physics up
to that point—had never been tested in the sort of
weak interactions that are responsible for the in-
stability of many elementary particles and atomic
nuclei. They proposed several experiments, and
when these were performed they demonstrated
that, in these interactions, the symmetry was indeed violated. This was a sensational discovery,
and all of us were gripped by its implications. In
the fall of 1957, Lee and Yang, who were both
at the Institute, were awarded the Nobel Prize for
Physics. Soon after my arrival at the Institute, I
studied the membership roster. There were several
physicists whose work I had studied and admired,
but I came across a name that really surprised me:
Reinhold Niebuhr.

Niebuhr, whom I had never met, was a hero
of mine. During the ten years I was at Harvard,
1947–1957, I had heard him preach and debate
several times. He was a great orator. His language
was simple but his ideas were profound and complex. He was also an imposing figure and gave the
impression that you were directly in his gaze. When
he spoke of the dangers of neurotic preoccupation

with self, I was sure that he was talking specifically to me. As it happened, my father, who was a rabbi, had been on several liberal commissions with Niebuhr—so I thought that if the occasion arose I would introduce myself to him. In the event, the occasion arose quite quickly. We all took our lunches in the Institute cafeteria, and I noticed Niebuhr and his wife eating alone. I passed by their table and said hello. They invited me to sit down. When Mrs. Niebuhr—Ursula—and I were alone for a minute, she told me that her husband had suffered a series of strokes, which was one of the reasons he had sought the relative tranquillity of the Institute. She said that he was feeling a little depressed and that if I would drop by their apartment from time to time it might help to cheer him up. I took advantage of this invitation. Niebuhr had sensed the excitement of the physicists and wanted me to explain what they were so exited about, which I did as best I could.

On one visit I noticed a book by W. H. Auden—I now forget which. Auden was another of my heroes and, as it happened, he was giving the Christian Gauss Seminars on literary criticism at Princeton University and I had attended some of his lectures. I mentioned this to the Niebuhrs, who were surprised and delighted. I will never forget how Ursula Niebuhr pronounced Auden's first name, Wystan. It sounded like "whistle"—with a hiss on the *s*. I did not know at the time that the Niebuhrs and Auden were close friends. Although she and Auden had been at Oxford at the same time—the late 1920s—they had not met until 1940. Niebuhr had become a sort of spiritual advisor to Auden, who dedicated his collection *Nones* to them. I felt that I had conveyed a useful piece of information to the Niebuhrs, and forgot about it.

Not long afterward, however, I was taking the "dinky"—the small train that connects Princeton to Princeton Junction—when on came Auden. You couldn't miss him. By that time, he had a face which he said looked like a wedding cake that had been left out in the rain. The train was pretty full, but there was an empty seat next to me and Auden took it. I told him that the Niebuhrs were in Princeton, which he had not known. He seemed pleased to learn this, and again I more or less forgot about it. But one morning, some days later, I was called to the hall telephone in our building. None of us had private phones in our offices, since Robert Oppenheimer, our director, had decided that these might distract us from our work. The hall phone distracted everyone from their work. It was a call from Oppenheimer's secretary. She said that I was expected for lunch and should come at lunchtime to Oppenheimer's office. I replied that there was surely some mistake, since there was no earthly reason Oppenheimer would want to have lunch with me. She said there was no mistake and I was expected.

Even as I write this, more than half a century later, I can still see the scene in Oppenheimer's office. There was Oppenheimer, impeccably dressed as usual, and his wife, Kitty. There was Sir Llewellyn Woodward, a noted British historian who had come from Oxford to the Institute, and his wife. There were the Niebuhrs, and Auden, and myself, a very minor postdoctoral. Whatever this was about, I am sure it reflected the fine "Italian hand" of Ursula Niebuhr, who wanted Oppenheimer and Auden to meet. After a few introductions, we all marched into the cafeteria.

A special table had been prepared for us in the center of the dining area. I am sure that we were served and did not stand in the cafeteria line. There was probably also wine. I remember the very fishy looks that my colleagues who were at our usual tables gave me. Freeman Dyson, who was then a professor at the Institute and a friend of mine, looked particularly amused. I had no opportunity to explain.

I wish I could tell you that the conversation was memorable. Oppenheimer, who was sitting across from Auden, seemed rather ill at ease. At one point he told Auden that he had studied Sanskrit at Berkeley in the 1930s. This did not make any impression, and Auden and Ursula Niebuhr then engaged in a lively conversation which pretty much ignored the rest of us. The Woodwards said nothing. Niebuhr caught my eye, and from the look on his face I would guess he was thinking, "This too shall pass." After lunch I took Auden to meet Dyson. They played some word games on Dyson's blackboard, and then Auden left and I never saw him again.

I have thought about this extraordinary occasion many times over the years, usually with a sense of regret. If only I had it to do again, I would have asked Auden which "English bishops on the quantum theory" he was referring to in his Byron poem. What a conversation that might have turned into. But, myself excepted, everyone who was at that table is now gone. "English bishops on the quantum theory"—How many can there have been? Finally I decided to try to find out. The first person I asked was Dyson. His answer was immediate: the Bishop of Birmingham, Ernest William Barnes. This suggestion was reinforced by the Auden scholars Edward Mendelson and Arthur Kirsch. But I have to confess that when Dyson said his name, I did not have the foggiest idea who Bishop Barnes was or why he would have had anything to say about the quantum theory. I am now going to tell you what I have learned.

Barnes was born on the first of April 1874—All Fools Day—in the town of Altrincham in Cheshire. His father, John (Starkie) Barnes, was a schoolmaster; and his mother, Jane (née Kerry) was the daughter of the village shoemaker at Charlbury in Oxfordshire. The family moved to Birmingham in 1876. I would be fascinated to know at what age and in what way Barnes began to show special mathematical ability. Mathematicians begin very young. To take an example, Dyson once told me that when he was still young enough to be "put down for naps," he invented for himself the notion of the convergent infinite series. He noticed that if you added $1 + \frac{1}{2} + \frac{1}{4} + \frac{1}{8} + \ldots$, the sum approached two. Where in this spectrum was Barnes? I do not know. In 1886, Barnes entered the King

Edward VI grammar school on a scholarship. Secondary education then was neither free nor compulsory. All of Barnes's higher education was done on scholarships. The family could not have afforded the tuition. At King Edward's, Barnes had the good fortune to come upon a truly great mathematics teacher, Rawdon Levett, who recognized Barnes's ability and could guide him into learning things like non-Euclidean geometry. Barnes was grateful to Levett for the rest of his life. It was Levett who encouraged him to apply for a fellowship to Trinity College in Cambridge, which he won.

To graduate with honors from Cambridge, students had to take the so-called Tripos examination. The mathematics Tripos consisted of hard problems that students had to complete in a certain time limit. The student who did best was labeled the Senior Wrangler. Where the term "wrangler" came from in this context no one seems to know. The student who did worst was known as the Wooden Spoon and was indeed given a spoon, along with third-class honors. As one might imagine, the list of Senior Wranglers from that period is quite impressive. It includes such people as the physicist Lord Rayleigh and the astronomer Arthur Eddington. But what one might not expect is that the list of Second Wranglers is even more impressive. It includes such people as James Clerk Maxwell, the greatest physicist of the nineteenth century, and William Thomson—Lord Kelvin—who was not far behind. Kelvin's case is instructive. In one of his courses, he had come in with a theorem and its proof. The theorem was on the Tripos, and by that time Kelvin had forgotten the proof. He spent a good deal of his exam time reconstructing it. The man who became the Senior Wrangler had it memorized. One wonders how Einstein would have done on the Tripos.

In 1896, Barnes was Second Wrangler. The Senior Wrangler was a New Zealand born physicist named Richard C. Maclaurin—hardly a household name, although he did become the president of MIT. But two years later Barnes won the First Smith's Prize for his Ph.D. thesis, an even greater honor which led to a fellowship at Trinity. For something like a decade Barnes did very significant work in pure mathematics. His work on the theory of functions is still being adumbrated. When I looked it up on the Web, I was interested to note that one of the papers cited was written by Dyson. In 1909, at the age of thirty-five, Barnes was elected a Fellow of the Royal Society. Soon after, he stopped doing creative mathematics. Dyson told me that G. H. Hardy, one of the most distinguished of the twentieth-century mathematicians, told him that once Barnes had been elected to the Royal Society he dropped mathematics like a "hot brick." It would certainly seem that way. Perhaps Barnes had realized that he would never be a mathematician in the class of people like Hardy and, above all, Ramanujan. Nobody was in the class of Srinivasa Ramanujan. He had been discovered in 1913 by Hardy, who invited him to Cambridge. He needed a formal tutor, and the more senior Barnes took on the responsibility. Barnes was remembered as a very good lecturer with a deep interest in mathematical education. Maybe it was not enough.

Barnes's family religious tradition was Baptist and Wesleyan, but the King Edward School was strictly Church of England. Barnes had no difficulty adapting and his parents did not object. In 1902 Barnes became ordained as a deacon in the church, and from that time on he preached sermons in various locales around Cambridge and elsewhere. From the beginning, his sermons had a scientific content. Confronted with the facts of evolution, for example, he could not take the Bible as a literal cosmogony. These sermons were referred to by others as Barnes's "gorilla" sermons. Another consistent theme of Barnes's sermons was pacifism. He was a confirmed pacifist from the time of his youth, and he never changed through two world wars. At the time of the First World War, this got him into trouble with his more conservative colleagues, but nothing like the trouble that Bertrand Russell got into. Russell was then at Cambridge, and his vivid denunciations of the war landed him in jail. Barnes was prepared to give Russell a character reference, even though they were not close friends and Barnes did not much approve of Russell's lifestyle. In 1915, Barnes received an offer he could not refuse. The Temple Church in London was looking for a new Master. The church, which was constructed by the Knights Templar in the twelfth century to serve their spiritual needs, had, in the twentieth, a very distinguished congregation. Barnes was only forty-three, young to be a Master of the Temple. It meant leaving Cambridge and moving to London. There was also a personal matter. Barnes had fallen in love with Adelaide Ward, the only daughter of a distinguished family. She had great hesitations about marriage—but when Barnes told her that unless she came to London with him he would refuse the job, she relented. It was a very happy marriage.

Barnes remained at the Temple until 1924. Then he received a letter from Prime Minister Ramsay MacDonald informing him that his name had been submitted to the king for an appointment as the Lord Bishop of Birmingham. The Church of England was the state religion and such appointments required royal approval, which Barnes got. It automatically carried a membership in the House of Lords. This enabled Barnes to participate in the debate in Parliament in 1927–1928 about revising the Book of Common Prayer, which could only be done with the consent of Parliament. Barnes opposed the revisions, and they were defeated by a narrow margin. When it came to church practice, Barnes was known to be very strict about not allowing any trace of Catholic ritual to enter into Anglican practice. And on the matter of transubstantiation, he was dogmatic. In a sermon he said, "I am quite prepared to believe in transubstantiation when I can find a person who will come to the chapel of my house and tell me correctly whether a piece of bread which I present to him has under-

gone the change for which believers in transubstantiation contend."[3] As one might imagine, this
sort of thing rendered Anglo-Catholics, of which
Auden's mother was one, livid. That Auden would
make a reference to Barnes—however concealed—
in his poem must tell us something about his relationship to his mother. In fact this was not Auden's
first mention of Barnes in his poetry. The earlier
one is much more explicit and curious.

In September 1931, the British Association for
the Advancement of Science had its centenary
meeting. It was a stellar gathering featuring such
events as General Jan Christiaan Smuts giving a
lecture on "holism" (his term). There were the
public scientists of the time, such as James Jeans.
Jeans had been a first-rate physicist, but by the
1930s he was writing popular books infused with
Christian mysticism. Barnes, who was also there,
must have found this appealing. The meeting must
have caught Auden's attention. There seem to be
echoes in his long poem *The Orators: An English
Study*, which he wrote soon after the meeting.[4] It
is a remarkable and strange poem—partly in prose
and partly in rhyming verse. Toward the end come
the lines,

Their day is over, they shall decorate the Zoo,
With Professor Jeans and Bishop Barnes at 2d a
 view,
Or be ducked in a gletcher, as they ought to be,
With the Simonites, the Mosleyites and the I.L.P.

"Their" refers to various people that Auden knew.
But what an odd association: Jeans and Barnes,
along with the Simonites, the Mosleyites, and the
Independent Labour Party.

The Gifford Lectures, well known among theolo-
gians and delivered every year at one or another
Scottish university thanks to a bequest of Adam
Lord Gifford (who died in 1887), were intended to
"promote and diffuse the Study of Natural Theology in the widest sense of the term . . . in other
words, the knowledge of God." "Natural Theology" meant theology consistent with the established principles of science. Over the years, the
Gifford lecturers have been an extraordinary group.
The scientists included people like Niels Bohr and
Werner Heisenberg. There were philosophers such
as Henri Bergson and Albert Schweitzer. Niebuhr
gave the 1938–1940 series. Dyson told me that
when he gave the 1984–1985 series in Aberdeen,
no one was allowed to ask questions. Apparently

you *were* allowed to laugh. In his biography of his father, John Barnes notes that when his father gave the 1927–1929 series, he provoked laughter by filling the blackboard with equations from Einstein's theory of general relativity. I would imagine that it was nervous laughter. The lecturers were expected to produce short books that would contain their lectures and that would be published not too long after them. Barnes's book, *Scientific Theory and Religion*, his 685-page magnum opus, was not published until 1933.[5] He spent every moment of his spare time in the intervening years preparing it.

The book covers all the natural sciences. There is rather little about mathematics. Perhaps Barnes did not think of it as a natural science. From time to time, he can't resist a dig. For example, he writes, "I have, personally, little doubt that biological research will in due course prove a human virgin birth is possible."[6] He adds,

> There are some who may shrink from any such investigation of the Virgin Birth as I have indicated. For my own part I am convinced that we must abandon the practice of arguing in vacuo, as it were, for such a mode of argument was the typical advice of medieval scholasticism. On the basis

of observed facts of Nature, and by arguments drawn from analogies as come within the range of our observations, we must approach all problems which present themselves. I would add that reverence and truth can always be combined, unless the object of our reverence happens to be untrue.[7]

Pace Auden's mother.

Barnes's chapters on the theory of relativity are really excellent. I doubt that many professional physicists could have written better ones. I was particularly interested in what he had to say about cosmology. The modern era in cosmology began when Einstein published a paper on the subject in 1917. This was a year after he had published his masterpiece on general relativity and gravitation. At the time, the "cosmos" consisted only of our own galaxy, and Einstein's concern was that his theory predicted that the gravitational attractions would make the whole thing collapse. So in his 1917 paper he added a kind of anti-gravitational term to his original equations to keep the universe stationary. Yet a few years later a Russian named Alexander Alexandrovich Friedman, completely unknown to Einstein, sent him a copy of a paper he was going to publish showing that, depending on how matter was distributed in the cosmos, there were solutions of Einstein's original equations in which the universe expanded or contracted. Einstein's first reaction was to claim that Friedman had made a mistake. But it was Einstein who had made a mistake. Then he claimed that Friedman's solutions were correct but irrelevant, since the universe was stationary. Here things stood until 1927.

At this time, a Belgian Catholic priest and astronomer named Georges Henri Joseph Edouard Lemaître published a paper in which he rediscovered Friedman's solutions. But he made a crucial next step. If the universe is expanding according to these equations, then every galaxy is receding from every other galaxy at a speed that increases as the distance of separation increases. We can think of the galaxies as being on the surface of a balloon that is being blown up. All the dots on the surface are receding from all the other dots. But Lemaître argued that this had an observable consequence. The spectral colors of the light from these galaxies should be shifted to the red end of the spectrum (Doppler shift) as the distance and hence the speed of recession increases. In 1929, the American astronomer Edwin Hubble published his observational results that showed that this indeed is what is happening. In his paper, he says that he is confirming the prediction of Lemaître. Einstein gave up his objections. But ironically, it now seems that the universe is expanding too rapidly to be accounted for by the Friedman equations, and that some "anti-gravity" is needed.

Barnes discusses the Lemaître model in detail. He concludes that there is something wrong. The time scales don't work out. If the universe were some millions of years old, the observed distances would have to increase too rapidly. But in fact cosmologists now think that the age of the universe is about 13.7 billion years, so the time scales do work out. Lemaître discusses where it all came from. He envisions a "cosmic egg" that exploded, and voilà —the first Big Bang model. But where in this is God? Here is what Barnes writes:

> Must we then postulate Divine intervention? Are we to bring in God to create the first current of Laplace's nebula or to let off the cosmic fire-work of Lemaître's imagination? I confess an unwillingness to bring God in this way upon the scene. The circumstances which thus seem to demand His presence are too remote and too obscure to afford me any true satisfaction. Men have thought to find God at the special creation of their own species, or active when mind or life first appeared on earth.

They have made him God of the gaps in human knowledge. To me the God of the trigger is as little satisfying as the God of the gaps. It is because throughout the physical Universe I find thought and plan and power that behind it I see God as the creator.[8]

This does not differ a great deal from Einstein's "Old One"—his God.

Barnes's chapter on the quantum theory, unlike the ones on relativity, seems a bit antiquated to me. To give an example: After the New Zealand–born physicist Ernest Rutherford discovered the atomic nucleus with his young collaborators Hans Geiger and Ernest Marsden, he proposed a model of matter. The nucleus had a positive electric charge to balance the negative charges of the surrounding electrons, so that the atom was electrically neutral. These positive charges were provided by the protons in the nucleus. But there had also had to be neutral objects of about the same mass as the protons, to make the masses of the nuclei come out. Rutherford made the natural assumption that these neutral objects were electrons and protons bound together. Indeed, when one of Rutherford's students, James Chadwick, made a direct observation of these neutral objects in 1932, he announced that this is what he had discovered. Already there were naysayers such as Wolfgang Pauli, who said that this must be wrong. But Chadwick dismissed them. He was wrong, and by 1934, when Barnes published his book, there was no doubt that these neutral objects—"neutrons," as they were named—were elementary particles in their own right. Barnes describes Chadwick's experiment, but he presents Rutherford's old model as if it were still valid. I find this very puzzling. Perhaps it reflects the fact that Barnes was not much in contact with contemporary physicists. This manifests itself in the rest of the chapter.

So far as I could see, there was nothing wrong in what he wrote. It is what he didn't write that struck me. He mentions almost in passing the work of Erwin Schrödinger—but then does not exhibit his equation, which is the heart of the matter. The mathematics is much simpler than that of general relativity, which he goes into in considerable detail. He also mentions in passing the work of Paul Dirac but never mentions Dirac's 1928 discovery of the equation that bears his name, which brought relativity and the quantum theory together and which predicted the existence of anti-particles. The first of these, the positron, was discovered in 1932, two years before Barnes published his book. He has a clear, brief discussion of the Heisenberg uncertainty principle, which Heisenberg presented in 1927. But there is no hint of the debate between Einstein and Bohr on the meaning of the theory—a debate which began about the same time. It is a strange chapter. I kept wondering: Did Auden read it? Was this his "bishop cheery" writing on the quantum theory? Perhaps Auden read one of the concluding paragraphs.

Old traditions embedded in Christian thought, describe a time when "the earth was without form and void": and probably most Christian philosophers in the past believed in a process of creation by God, in time and "out of nothing." We seem, in the analysis of matter to which Einstein's general relativity leads, to see "in the beginning" a process by which form or structure was given to the void of space-time. As the many complex forms which then arose assumed an ever greater complexity, the material world took shape. It is natural to ask whether, in such development, there was creative activity, the emergence of something new. I feel constrained to answer in the affirmative. Things were other and more diverse at the end than at the beginning. Even though the process of assuming greater complexity may be exhibited by a mechanical sequence, it may none the less conceal or embody genuine creative activity. . . . Did not such complexity involve in its making a series of acts rightly to be called creative, or at least a series of changes which led to the creation of something that did not previously exist?[9]

As for Barnes, in 1947 he published a very controversial book, *The Rise of Christianity*, in which he expressed his rationalistic views fully. It created a furor and provoked attempts, unsuccessful, to make him resign. He finally did, in February 1953. He died in November of that year.

In 1940, the chocolate magnate Edward Cadbury established a chair in theology at the University of Birmingham. Niebuhr was in the process of giving his Gifford Lectures and Barnes proposed him as the first occupant of the chair. The suggestion was turned down. But suppose it had not been. Then the Niebuhrs would not have met Auden and I would not have met any of them. The encounter that inspired me to make this inquiry in the first place would never have happened.

:: 2 ::

Quantum Buddhists

On the Quantum Cruise you will directly experience your connection with the Unified Field through music, movement, art, sound, drumming, Quantum Imagery, and other dynamic, fun and transformative processes. Integrate your intellectual understandings into the very molecules of your physical body.

—Quantum Cruise website, accessed January 10, 2006

My confidence in venturing into science lies in my basic belief that as in science so in Buddhism understanding the nature of reality is pursued by means of critical investigation: if scientific analysis were conclusively to demonstrate certain claims in Buddhism to be false, then we must accept the findings of science and abandon those claims.

—The Dalai Lama[1]

When one listens to descriptions of subatomic particles such as quarks and leptons in modern physics, it is evident that the early Buddhist atomic theories and their conception of the smallest indivisible particles of matter are at best crude models. However, the basic thrust of the Buddhist theorists, that even the smallest constituents of matter must be understood as composites, appears to have been on the right track.

—The Dalai Lama[2]

ON AUGUST 30, 1983, the Dalai Lama and a delegation of Tibetan monks visited CERN. "CERN" is an acronym for Conseil Européen pour la Recherche Nucléaire, although in the official name "Organisation" has now replaced "Conseil." CERN, which is on the border between France and Switzerland just outside Geneva, is the largest laboratory in the world devoted to the physics of elementary particles. It is an international organization in every sense, although I am not aware that any Tibetans have ever worked there, at least as scientists. Why the Dalai Lama chose to make this visit is somewhat mysterious to me, but accounts of his boyhood do describe his interest in taking apart and putting together machines to see how they worked. In any event, on this occasion there were something like thirteen monks. CERN, which is used to receiving delegations, arranged a formal lunch for the Tibetans and a corresponding number of people from the laboratory. One of them was my friend John Bell, who died suddenly of a cerebral hemorrhage in the fall of 1990 at the age of sixty-two. But some months before his death I had the chance to discuss the Dalai Lama's visit at some length with him.

The reason for Bell's having been part of the CERN delegation is quite clear. Bell had almost single-handedly renewed interest among physicists in the foundations of the quantum theory. Until Bell's work, a few decades before the monks' visit, this subject was pretty much the province of elderly physicists such as Bohr and Einstein, with very few others paying much attention. The Einstein-Bohr debates did not seem to lead anywhere. Bell's work, which I will discuss more fully later, made the subject an experimental science. It received wide publicity, followed by considerable misunderstanding, which I will also discuss later. There was, and is, an attempt to link it with Eastern religions such as Buddhism, and possibly this was one of the motivations that led the Dalai Lama to visit CERN. In any event it was a "state visit," during which the Dalai Lama, who speaks English both fluently and charmingly, would speak only through an interpreter. Bell recalled some of the discussion.

At one point the Dalai Lama asked: If the elementary particles were point-like, how could they add up to make extended objects? "It did not seem to occur to him that there could be spaces between the points—very odd," Bell remarked. Bell was interested in exploring with the Dalai Lama the notion of the Big Bang. He cautioned that the Big Bang was only a fashion in modern science, and that fashions could change. But could it be reconciled with the Buddhist canon according to which the universe is constantly recycled? The Dalai Lama said that this was not one of the precepts to which they were completely committed; they would have to study the scriptures carefully. Then he added, "There is some room for maneuver." I was so struck by the "some room for maneuver" that I asked Bell whether this was his turn of phrase or that of the Dalai Lama. Bell said "room for maneuver" was exactly what the interpreter had said, and Bell found that wonderful. In fact, although Bell did not tell me this—he may have forgotten or never knew—this session was filmed and one can see Bell asking this question and the Dalai Lama answering.[3]

Bell brought up the question which is often asked: Why did science not take root in countries like China with a Buddhist tradition?[4] China has a long history of inventions—many of them of great significance. But nowhere do you find a Galileo, a Newton, a Darwin, to say nothing of an Einstein. There must certainly have been potential scientific genius, but there was no context. A tiny country like Denmark could produce a Bohr, because once such a person appeared, there was a tradition into which he could be fitted. Mathematics is a little different. A genius can almost make his own context. Ramanujan, for example, had some mathematical education, but when he went to Cambridge to work with G. H. Hardy, he did not know what a proof was. He simply produced fantastic results out of his imagination that he knew were true. "Buddhism is an inward religion concerned with individual salvation," noted Bell. "The study of how stones accelerate must have seemed unimportant. The Tibetans were very hazy about modern physics, and I had the impression that they did not consider these analogies significant. Physics is ephemeral and rapidly changing and applies to only a limited set of phenomena. It would be absurd to anchor your religious beliefs to theories of physics. I did not have the impression that they were looking around and grasping for straws. They seemed to be quite comfortable in their own traditions."

I took advantage of my interviews with Bell to
ask him about another encounter he had had with
Eastern religious mysticism. This one took place in
1979 at a conference with the Maharishi Mahesh
Yogi, who died in February 2008. As it happens,
the Maharishi had been a physics major at the University of Allahabad in the North Indian state of
Uttar Pradesh. In 1939, after his graduation, he
became a disciple of Swami Brahmananda Saras-
wati who gave him the name Bal Brahmacharya
Mahesh. Not being a Brahmin, the Maharishi could
not succeed his master after the latter's death in
1953. So he struck out on his own, teaching Tran-
scendental Meditation to various disciples, including most notably the Beatles. In the early 1970s
he founded the eponymous Maharishi European
Research University in what had been an elegant
resort hotel near Weggis, above Lake Lucerne in
Switzerland. The hotel had a large circular dome
which bore inscriptions such as "Center of World
Government." There was a Minister of Health and
Immortality. He appointed as chancellor a physicist named Larry Domash.

A brief Google search turned up a 1973 paper
entitled "Interactions among Multiple Lines in the
8446-a Atomic-Oxygen Laser," published in the

Physical Review (our refereed trade journal) and coauthored by Domash. The byline indicates that, when the paper was written, he was employed by the NASA Electronics Research Center, in Cambridge, Massachusetts. But there is a footnote that reads "Present address, Maharishi International University, Rishikesh, India." It was surely Domash who selected the list of invitees for the 1979 conference. In addition to Bell, there were other well-known quantum theorists. "The Maharishi," Bell recalled, "was a much more regal figure than the Dalai Lama. He sat on a sort of throne dressed in white robes. At his feet were acolytes—young women—also dressed in white robes. In the middle of this gathering were the invitees. It was a very uncomfortable situation for a scientist. I made a speech that was skeptical in character. Domash was trying to see some analogy between the state which people reach in meditation and the ground state of a superconductor." A superconductor is a quantum mechanical state that some materials attain when they are cooled to very low temperatures. As the name implies, in this state electricity is conducted with no resistance. I have read claims by Domash that neurons are superconductors. Bell thought that all of this was nonsense and said so.

They asked the Maharishi for his views. I was shocked to learn that the Maharishi claimed he could make rain. First you see a blue sky and a little cloud, and then you relax. The cloud grows and there is rain. Unfortunately this was not something you could order, especially with a skeptical audience. At that time, he had a class whom he was trying to teach to levitate above the ground. He was interested in all the queer aspects of quantum mechanics which he thought had some connection to Eastern mysticism. I did not have a chance to converse with him. For me, he was just a figure on the throne making pronouncements. But I liked the Maharishi setup. The meals were good. It was vegetarian.

Bell had been a vegetarian since the age of sixteen.

While Bell expressed a good deal of skepticism about the Maharishi, his feelings for the Dalai Lama were quite different. Knowing John Bell as I did, I am sure that his description of his meeting with the Dalai Lama is an accurate representation of what occurred. In fact, it is borne out by the film of that session. The pity is that Bell and the rest of the CERN people did not know what the Dalai Lama's scientific background was even at the time of his visit in 1983. If they had, or if Bell had had a chance to talk to the Dalai Lama privately and informally, I think the results would have been very different and the impressions created would have been very different. As it was, Bell gave a short, but brilliant, introduction to the quantum theory. There was an attempt at dialogue, which foundered because of all the people present. In 2005, the Dalai Lama published a small book entitled *The Universe in a Single Atom: The Convergence of Science and Spirituality*. In the book, he makes it clear that his interest in science goes back to his early boyhood. The Thirteenth Dalai Lama—the "previous body"—died in 1933. The search for his reincarnate did not begin for four years. In 1937 a search party from Lhasa found a two-year-old boy in Amdo—a province in the northeast of Tibet—whom they recognized as the incarnate—the "present body," as the Fourteenth Dalai Lama refers to himself. They had to pay a large bribe to the Chinese, who claimed sovereignty, to allow the child to leave. Once in Lhasa, he was installed in the Potala Palace and his education in Buddhism began. Science played no part in it—I am not even sure he was taught basic geometry.

The Potala is a giant warren which has, it is said, a thousand rooms. It is now—as I can testify, having first visited it in 1987—a museum in which the Chinese extract as much money as they can from tourists. Tibetans can't afford to go there. Among the exhibits are the stupas, the gold-encrusted tombs in which all the Dalai Lamas from the Fifth to the Thirteenth lie in state. The Thirteenth was an interesting man. He tried without success to modernize Tibet. During the early years of his regime, there was a power struggle between the Chinese and the British, as both claimed dominion over Tibet. The Chinese invaded in 1910, and the Dalai Lama fled to Sikkim, a Himalayan British protectorate. He used the same route that his successor would use in 1959. The political officer in charge of Sikkim at the time was a man named Charles Bell—no relation to John.

Charles Bell and the Thirteenth Dalai Lama had an amiable relationship, and Bell gave him several "souvenirs" of his visit. There was a telescope and a hand-wound pocket watch which gave the time in various time zones. There were also three automobiles acquired later—two 1927 Baby Austins and a 1931 American Dodge. There was—and is—no motorable road from Sikkim to Lhasa so the automobiles had to be dismantled and the parts carried over the Himalayas by porters, human and animal. They were reassembled and maintained,

but how much they were driven around the city—
the only place in the country that had anything like
roads—is unclear. The young Fourteenth Dalai
Lama, while exploring the Potala, discovered these
treasures. He managed to take a short drive in one
of the cars, and had a small accident in which a
headlight was damaged. The watch he learned to
take apart and put back together. No one explained
to him the use of the telescope, but he figured that
out too. He began observing the Moon and noticed
that there were shadows being cast by various pro-
tuberances. He called in his tutors, who did not
understand what they were seeing. The Dalai Lama
explained it must mean that the Moon was being
illuminated by the Sun—a scientific discovery. Using his telescope, he made another discovery: there
was a European in Lhasa. This turned out to be
the Austrian mountain climber Heinrich Harrer.

In 1939, Harrer was part of a German expedition to climb Nanga Parbat. This is the ninth-
highest mountain in the world and one of the most
difficult to climb. It is located in Pakistan, which
was then part of British India. When the Second
World War broke out, the climbers were interned
as enemy aliens by the British. It is very unlikely
that the British knew that Harrer was in the SS and

had been photographed with Hitler. He was a German hero for having climbed the north face of the Eiger in Switzerland. The Dalai Lama certainly did not know that Harrer had been in the SS. This was revealed publicly only in 1997. Harrer's attitude was that bygones should be bygones. He died in 2006 at the age of ninety-three.

Harrer—who escaped from the British internment camp in 1944, and arrived in Lhasa two years later after a harrowing trek across Tibet—was the Dalai Lama's first contact with the outside world. He repaired the movie projector, and the two of them watched such films as Lawrence Olivier's *Henry V*, which had been imported from India. Harrer also gave the Dalai Lama his first lessons in world geography. Harrer had been accompanied on his escape by another Austrian mountaineer named Peter Aufschnaiter, who was actually the leader of the Nanga Parbat expedition. While in Tibet, Aufschnaiter had mapped very remote areas of the country for the government and helped design the hydroelectric power plant in Lhasa. After the Chinese invasion in 1950, both men left the country, Harrer for Austria and Aufschnaiter for Nepal, where he married a Tibetan woman. I met Aufschnaiter and his wife when I was in Nepal in 1967. He was doing agricultural engineering for various international organizations. He died in 1973. In 1953, Harrer published his much-admired *Seven Years in Tibet,* which was followed by Aufschnaiter's *Eight Years in Tibet.*

In October 1950, the Chinese army invaded Tibet. Not long afterwards the Dalai Lama, at age sixteen, was invested with the full powers of his office, something which ordinarily would not have happened until he was eighteen. During this period, he visited China and met Mao. In his book, the Dalai Lama says that this was the first time he had ever seen a real city, with all its technological variety. He also visited India, where he was to flee in 1959.

It was in 1973 that he made his first visit to the West. In England—he does not tell us how—he met the Austrian-born philosopher of science Karl Popper. He and Popper shared the experience of having been exiled forcibly from their native countries. The Dalai Lama says that his English was not yet good enough to get the full benefit of his conversations with Popper. Popper's main theme in the philosophy of science was the notion of "falsifiabil-

ity." No number of confirmations suffices to prove
a theory, but one failed prediction suffices to dis-
prove it. The Dalai Lama writes,

> Another of the differences between science and
> Buddhism as I see them lies in what constitutes a
> valid hypothesis. Here too Popper's delineation of
> the scope of a strictly scientific question represents
> a great insight. This is the Popperian falsifiability
> thesis, which states that any scientific theory must
> contain within it conditions under which it can be
> shown to be false. For example the theory that
> God created the world can never be a scientific one
> because it cannot contain an explanation of the
> conditions under which the theory could be proven
> false. If we take this criterion seriously, then many
> questions that pertain to our human existence,
> such as ethics, aesthetics, and spirituality, remain
> outside the domain of science. By contrast, the do-
> main of inquiry in Buddhism is not limited to the
> objective. It also encompasses the subjective world
> of experience, as well as the question of values. In
> other words, science deals with empirical facts but
> not with metaphysics and ethics, whereas for Bud-
> dhism, critical inquiry into all three is essential.[5]

The Dalai Lama's second visit to Europe was in
1979. Keep in mind in what follows that this was
four years before his visit to CERN. It was during

his 1979 visit that the Dalai Lama began his studies of the quantum theory. In a way that he does not explain in his book, he acquired two "tutors," Carl Friedrich von Weizsäcker and David Bohm. Two more disparate figures it is impossible to imagine. Weizsäcker, who was born in Kiel in 1912, came from an aristocratic German family. He had inherited the title of Freiherr (baron). His father, Ernst, served as secretary of state, became a high official in the SS, and was convicted of war crimes at Nuremberg. His brother Richard became president of Germany.

C. F. von Weizsäcker studied physics with various people, including Niels Bohr and Werner Heisenberg. Heisenberg gave him the problem of working out the detailed optics of what is called the "Heisenberg microscope." This is like any other microscope, except that Heisenberg used it in a "thought experiment," to determine the position and momentum of a subatomic particle. The idea is that a quantum of radiation (many quanta, in a realistic situation) is scattered off the particle and arrives at the microscope lens. The resolving power of the microscope—the capacity to determine the particle's position—is proportional to the wavelength of the quantum. But each quantum carries a momentum that is inversely proportional to the wavelength. So the more precisely one measures the position with the microscope, the more momentum one imparts to the particle. This is a textbook example of the Heisenberg uncertainty principle. But Heisenberg asked Weizsäcker to work out the details in a realistic case. After taking his Ph.D., Weizsäcker went to Berlin, where he became an assistant to the Austrian-born physicist Lise Meitner.

While he was with Meitner, he created what is known as the "Weizsäcker semi-empirical mass formula." This formula enables one to plot the masses of all the nuclei—except for the very lightest ones—in terms of a small number of empirically determined parameters. Meitner made use of this formula on the famous walk in the woods she took in December 1938 with her nephew Otto Frisch. She was able to argue that the experiments of her former Berlin colleagues Otto Hahn and Fritz Strassmann showed that the uranium nucleus had fissioned. Weizsäcker also proposed a process by which the Sun generates its energy. He did not work out the details. This was done by Hans Bethe, who had independently proposed the same process. When, in the fall of 1939, the German army drafted a number of scientists to work on nuclear energy, including the prospects for nuclear weapons, Weizsäcker and Heisenberg were among them. One of Weizsäcker's contributions in the summer of 1940 was a proposal to use what was later called plutonium in a nuclear weapon. In 1945, he was one of the German scientists who was captured and interned at Farm Hall, near Cambridge, England.

Weizsäcker, and Heisenberg for that matter, were "non-Nazis"—a term that was used at Farm Hall. Neither of them joined the Nazi Party. On the other hand, Heisenberg at the outset certainly wanted the Germans to win the war, and both of them, despite what they said later, were quite willing to work on an atomic bomb for Hitler. When Strasbourg (Strassburg in German) fell to the Germans, the university was taken over and Aryanized. Weizsäcker had no problem accepting a professorship there in 1942. His tenure was not long, as the city fell to the Allies in 1944. After the Germans were released from Farm Hall, Weizsäcker returned to Germany and played an important role in resurrecting German academic science. He was also an outspoken opponent of nuclear armament for Germany. His days as a creative scientist were over, but he wrote a good deal about philosophy and in particular about quantum mechanics. In quantum mechanics he was a Bohr positivist. I will explain this more fully later, but in essence he agreed with Bohr that there was no hidden classical substratum underlying the quantum theory. The theory, with all its probabilities and uncertainties, had to be accepted for what it was. Weizsäcker died on April 28, 2007.

David Bohm was the absolute antipode of Weizsäcker, which makes the Dalai Lama's choice of these two men as his tutors so remarkable. Bohm was born in Wilkes Barre, Pennsylvania, in 1917. His mother and father were Jewish immigrants from eastern Europe. His father owned a furniture store and was the assistant to the local rabbi. After graduating from Penn State, Bohm went to California, where Robert Oppenheimer served as his thesis advisor. Bohm became very involved with left-wing political organizations, including the Young Communist League. When the war broke out, Bohm could not get clearance to go to Los Alamos with Oppenheimer. Considering the people who were cleared—including Oppenheimer's brother, who had been a member of the Communist Party—this is rather odd. Bohm remained in Berkeley, where he did work connected with the atomic bomb project. After the war he joined the faculty of Princeton University.

In the early 1950s two things happened to Bohm which changed his life. First, he was called to testify before the House Un-American Activities Committee, where he took the Fifth Amendment. Princeton suspended him, and Oppenheimer suggested that he leave the country—rather strange advice, considering Oppenheimer's own situation, with his loss of his security clearance and his refusal to leave the country. Yet Bohm took this advice, going first to Brazil, then to Israel, then to England, where he died in 1992.

The second thing that happened was that in 1951 Bohm published his text *Quantum Theory*.[6] At the time, this was the only text I knew of where the conceptual foundations of the theory were discussed at length. People who knew Bohm then have told me that arriving at the position he took in the text required some personal struggle. To some Marxist-Leninist dialecticians, the Copenhagen interpretation was an anathema. For example, according to this interpretation an electron is neither a particle nor a wave. Depending on the experiment, it can manifest itself as one or the other or both. To Niels Bohr, these were "complementary" descriptions that reflected the limitations of language. Our language is classical and we do our best to apply it to quantum mechanics. But in some hands this can morph into the notion that we create reality: the world is a will and an idea. It is this idealism that the school of Soviet dialecticians objected to, and that apparently troubled Bohm when he first learned the theory. But his textbook reflects none of this. It is orthodox Bohr.

Bohm was quite surprised when, unexpectedly, Einstein called him. Einstein told Bohm that his book was the best attempt to refute his views that he had ever seen. Einstein believed that the quantum mechanical description was not complete, while Bohm argued in his book that it was. Einstein accepted the consequences of the uncertainty principle, but thought that this and the other aspects of the theory would emerge as limits of some sort of classical theory—which he never succeeded in creating. He invited Bohm to come and visit him to discuss these matters. Exactly what happened in this discussion I have never been able to find out. I once wrote Bohm to ask him. He answered my letter but not the question. It seems Bohm was persuaded that the orthodox interpretation of the theory—the one he had so brilliantly presented in his book—left something to be desired. It also seems that Einstein proposed something fairly specific, because he later expressed his disappointment that Bohm had not carried it out.[7] What Bohm did do was to produce a theory of "hidden variables," which was completely deterministic and classical but which reproduced all the results of the quantum theory, so long as effects of the theory of relativity could be set aside.

The Dalai Lama, if we can judge from his book, was very fond of Bohm. Like Popper, and himself, Bohm had been exiled from his home country. The Dalai Lama writes, "I particularly admired Bohm's extraordinary openness to all areas of human experience, not only in the material world of his professional discipline but in all aspects of subjectivity, including the question of consciousness. In our conversations, I felt the presence of a great scientific mind which was prepared to acknowledge the value of observations and insights from other modes of knowledge than the objective scientific."[8]

Later in the book he writes about Bohm's notion of "implicate order," which is to replace reductionism. This is a kind of holistic "flow"—perhaps comparable to the Buddhist concept of "wholeness." The Dalai Lama remarks,

I notice that there is a group of scientists and philosophers who appear to believe that scientific thinking derived from quantum physics could provide an explanation of consciousness. I remember having some conversations with David Bohm about his idea of an "implicate order," in which both matter and consciousness manifest according to the same principles. Because of their shared nature, he contends, it is not surprising that we find a great similarity of order between thought and matter. Although I never fully understood Bohm's theory of consciousness, his emphasis on a holistic understanding of reality—including both mind and matter—suggests an avenue by which to look for a comprehensive understanding of the world.[9]

In his book, the Dalai Lama presents his understanding of the problems in the interpretation of the quantum theory. It comes in a long and I think very remarkable statement, which I will quote in full and then comment on. He writes,

In physics, the deeply interdependent nature of reality has been brought into focus by the so-called EPR paradox—named after its creators, Albert Einstein, Boris Podolsky, and Nathan Rosen—which was originally formulated to challenge quantum mechanics. Say a pair of particles is created[10] and then separates, moving away from each other in opposite directions—perhaps to greatly distant locations, for example, Dharamsala, where I live, and, say, New York. One of the properties of this pair of particles is that their spin must be in opposite directions—so that one is "up" and the other will be found to be "down." According to quantum mechanics, the correlation of measurements (for example, when one is up, then the other is down) must exist even though the individual attributes are not determined until the experimenters measure one of the particles, let us say in New York. At that point the one in New York will acquire a value—let us say up—in which case the other particle must simultaneously become down. These determinations of up and down are instantaneous, even for the particle in Dharamsala, which has not itself been measured. Despite their separation, the two particles appear as an entangled entity. There seems, according to quantum mechanics, to be a startling and profound interconnectedness at the heart of physics.

The Dalai Lama goes on,

Once at a public talk in Germany, I drew attention to the growing trend among serious scientists of taking the insights of the world's contemplative traditions into account. I spoke about the meeting ground between my own Buddhist tradition and modern science—especially in the Buddhist arguments for the relativity of time and for rejecting any notion of essentialism. Then I noticed von Weizsäcker in the audience, and when I described my debt to him for what little understanding of quantum mechanics I possess, he graciously commented that if his own teacher Werner Heisenberg had been present, he would have been excited to hear of the clear, resonant parallels between Buddhist philosophy and his scientific thoughts.

The Dalai Lama concludes,

> Another significant set of issues in quantum mechanics concerns the question of measurement. I gather that, in fact, there is an entire area of research dedicated to this matter. Many scientists say that the act of measurement causes a "collapse" of either the wave or the particle function, depending on the system of measurement used in the experiment; only upon measurement does the potential become actual. Yet we live in a world of everyday objects. So the question is, how, from the point of view of physics do we reconcile our commonsense notions of an everyday world of objects on the one hand and the bizarre world of quantum mechanics on the other? Can these two perspectives be reconciled at all? Are we condemned to live with what is apparently a schizophrenic view of the world?[11]

As I was reading this, I could not help thinking of John Bell. It was his analysis of the Bohm version of the EPR experiment, the topic of the Dalai Lama's discussion, which showed that hidden-variable versions of the quantum theory would inevitably involve transmission of influences at speeds greater than that of light, making them at least on their face difficult to reconcile with relativity. The questions about how the quantum and the classical worlds coexist were the ones that bothered Bell the most. None of the proposed solutions entirely satisfied him. In Bohm's model, the world is classical but nonlocal. In other models, the world is entirely quantum mechanical, which requires explaining where everyday classical physics fits in. There are researchers who believe that the world is divided into patches, some of which are quantum mechanical and some of which are classical—which leaves the question of characterizing the patches. If only Bell and the Dalai Lama had gone off together to discuss these things by themselves. Just imagine what they could have taught each other.

Léon Rosenfeld

Dear $\sqrt{(\text{Trotsky} \times \text{Bohr})} = \text{Rosenfeld}$

—Wolfgang Pauli[1]

Is it not delightful that poor Bohr's only support-
ers in Paris should be this logical couple [de Broglie and Destouches], while all the youth are up
in arms against him "under the banner of Marxism"? Poor Marx too, I would add, since I belong,
as you know, to the almost extinct species of genuine Marxists: the kind of theology dished up under this name today is just as repulsive to me as to
you, perhaps even more so because I see it against
a background of what Marx really meant.

—Léon Rosenfeld to Wolfgang Pauli[2]

NIELS BOHR HAD great difficulty in actually writing a paper, or in giving a lecture for that matter.
One would be tempted to say that it was some curious form of dyslexia. This was especially true in his
mature years, as his output became more and more
philosophical—less mathematical. To deal with the

problem, Bohr hired assistants, part of whose job was to act as sounding boards and amanuenses. When none of these assistants were available, he pressed into service anyone who was at hand and who seemed qualified. One of these people was the late Abraham Pais. He wrote about this experience in his biography of Bohr.

> Bohr devoted tremendous effort and care to the composition of his articles. However, to perform the physical act of writing, pen or chalk in hand, was almost alien to him. He preferred to dictate. On one of the few occasions that I actually did see him write himself, Bohr performed the most remarkable act of calligraphy I shall ever witness.
>
> . . . As we were discussing the address that Bohr was to give on the occasion of the tercentenary celebration of Newton's birth, Bohr stood in front of the blackboard (wherever he dwelt, a blackboard was never far) and wrote down some general themes to be discussed. One of them had to do with the harmony of something or other. So Bohr wrote down the word "harmony."[3]

In his book, Pais gives what he recalls as Bohr's written version of the word "harmony." It is a squiggle that could be anything. Pais goes on, "However, as the discussion progressed, Bohr became dissatisfied with the use of harmony. He walked around restlessly. Then he stopped, and his face lit up. 'Now I've got it. We must change harmony to uniformity.'[4] Bohr then picked up the chalk and added to the previous squiggle a single dot, and then he put down the chalk with a look of great satisfaction. As I can attest, Bohr as a lecturer was something of a disaster. Not only was he often inaudible but he would do things like the following (which Pais also describes). In the course of a lecture, Bohr made a point and then changed his mind. He said "but"—and was silent until whatever followed the "but" played out in his mind. Then he resumed the lecture as if he had spoken his thought, never actually doing so. Over the years, Bohr had a number of assistants and collaborators. One of the most interesting and long-lasting was the Belgian physicist Léon Rosenfeld. They collaborated for more than thirty years, from 1930 until Bohr's death in 1962.

Rosenfeld was born on the fourteenth of August 1904 in Charleroi, Belgium, making him a couple of years younger than Pauli and Heisenberg. His father was an electrical engineer who died in an accident when Rosenfeld was fourteen. After high school he entered the University of Liège, graduat-

ing in 1926 with high distinction in mathematics
and physics. He then won a fellowship that enabled
him to study in Paris, at the Ecole Normale Su-
périeure. Among his teachers was the physicist
Paul Langevin, who had strong left-wing views
and eventually, in 1944, joined the Communist
Party. He was one of the Paris scientists who in the
1920s were actively engaged with social problems.
Rosenfeld took part, and this was the beginning of
his life-long preoccupation with Marxism. So far as
I can tell, Rosenfeld never joined the party. He did
visit Russia in 1934 with Bohr. His Marxism pretty
much stopped with Marx and Engels. Lenin had
already strayed from the True Belief, and Stalin
was beyond the pale.

Later on, I will have much more to say about
Rosenfeld, Marxism, and the struggle for the soul
of the quantum theory, but here let me say a
few words about Lenin's book *Materialism and
Empirio-criticism*, which he wrote while in exile
from Russia, in London and Geneva, in 1908. Although Rosenfeld does not refer to this book in any
of his writings that I have seen, it became the basis
of the Soviet Communist ideological canon with regard to science. Its echoes are apparent in the de-

bate, often vitriolic, that took place beginning in the 1940s over the interpretation of the quantum theory.

In some ways, reading Lenin's book is like watching one of those old black-and-white films in which comic actors throw pies at each other. Opponents "cravenly hide" or talk "gibberish." A typical sentence reads, "The Russian Machists will soon be like fashion-lovers who are moved to ecstasy over a hat which has already been discarded by the bourgeois philosophers of Europe."[5] Machists are variously "muddled," "brainless," and show "appalling ignorance." All good clean fun until it became possible in the 1930s to send the party's enemies—or anyone who disagreed—to the Gulag. Why Mach?

Ernst Mach was an Austrian physicist and philosopher of science who was prominent at the end of the nineteenth century and the beginning of the twentieth. His book *The Science of Mechanics* helped to shape the thinking of the young Einstein. Einstein was especially struck by Mach's polemical attack on Newton's idea of an absolute space and time beyond the reach of ordinary clocks and rulers. Mach believed that scientific theories were simply the economical description of observations.

He no doubt subscribed to the motto of Bishop George Berkeley, the eighteenth-century divine who famously stated, *Esse est percipi:* "To be is to be perceived." Lenin states in his book several times that all Mach did was to plagiarize Berkeley. Mach was an anti-atomist. He once had an exchange with his University of Vienna colleague Ludwig Boltzmann, a very great physicist to whom we owe, among other things, our notion of entropy in terms of probability. Mach said that he did not believe in atoms. When Boltzmann objected, Mach asked, "Have you seen one?" Lenin would have emphasized the "you" in Mach's question: "Have *you* seen one?" The point would have been that only your personal observation could enable you to say that atoms exist. For Lenin this was a slippery slope, at the bottom of which is the notion that there is no objective world outside yourself. Here, as we shall see, are the seeds of the objection that some Soviet scientists and philosophers raised concerning Bohr's interpretation of the quantum theory: his ideas could lead to the conclusion that to be is to be measured.

Interestingly, Einstein is never mentioned in Lenin's book. But one of my teachers, the noted philosopher of science Philipp Frank, gets a dig.

Lenin writes, "'The French mathematician Henri Poincaré,' we read in the work of the Kantian Philipp Frank, 'holds the point of view that many of the most general laws of theoretical natural science (e.g., the law of inertia, the law of conservation of energy, etc. . . .) of which it is often difficult to say whether they are empirical or of *a priori* origin are, in fact, neither one nor the other, but are purely conventional propositions depending on human discretion.'"[6] Professor Frank used to cite this with a good deal of amusement, especially at the notion that he, one of the founders of the Vienna Circle, whose members were all logical positivists, was a "Kantian." Now back to Rosenfeld.

After his Paris sojourn, Rosenfeld became an assistant to Max Born in Göttingen for the period 1927–1929. This was just after the creation of quantum mechanics, in which Born played an essential role. I would imagine that this must have been an incredibly interesting time for a young physicist—and a little discouraging, too, when one saw that people such as Heisenberg, Pauli, and Paul Dirac, who were about the same age, were making such profound contributions. I think it is fair to say that Rosenfeld never did physics at this level. Probably the most important physics paper he ever wrote was the one he published with Bohr in 1933. He had gone to Copenhagen in 1930. This paper had to do with whether, in the light of quantum mechanics and its uncertainties, the field strengths of electric and magnetic fields can be measured. Bohr and Rosenfeld showed that, with quantum mechanics, it is impossible to measure these fields at a single point (in contrast to the classical case); one can do so only over a small region of space. If one does this carefully, the Heisenberg uncertainty relations are respected. Rosenfeld must have impressed Bohr on a personal level, because he soon became Bohr's sounding board.

We get a glimpse of how this must have been if we look at Rosenfeld's account of Bohr's reaction to the 1935 paper written by Einstein, Boris Podolsky, and Nathan Rosen (the EPR paper discussed by the Dalai Lama). By this time, Einstein had settled in Princeton and Podolsky and Rosen were working with him there. The article was written in English, largely by Podolsky, and was entitled, "Can Quantum-Mechanical Description of Physical Reality Be Considered Complete?" Their answer was no, and they illustrated this with a "thought experiment" involving the measurement of the positions and momenta of widely separated particles. This is not the place to go into their argument, except as an illustration of the Bohr-Rosenfeld collaboration. It is nicely described by Pais in his biography of Bohr.

"This onslaught came down upon us as a bolt from the blue," [said Rosenfeld]. . . . "As soon as Bohr heard my report of Einstein's argument everything else was abandoned, we had to clear up such a misunderstanding at once." Rosenfeld has told me that Bohr was infuriated when he first was informed. The next day he appeared at his office all smiles, turned to Rosenfeld and said, "Podolski, Opodolski, Iopodolski, Siopodolski, Asiopodolski, Rasiopodolski." Poor Rosenfeld was bewildered. Bohr explained he was quoting from a Holberg play in which a servant suddenly starts talking gibberish. . . . When Rosenfeld remarked that Bohr had evidently calmed down, Bohr replied, "That's a sign that we are beginning to understand the problem. . . . They do it smartly but what counts is to do it right."[7]

Another instance occurred in the winter of 1939. After Otto Frisch had spent his Christmas vacation in Sweden with his aunt Lise Meitner, he returned to Copenhagen, where he was then working. He immediately told Bohr about their analysis of Otto Hahn and Fritz Strassmann's experiments on uranium. Meitner and Frisch had realized—as Hahn and Strassmann had not—that fission had occurred in the experiments. Bohr's reaction was the same as that of every physicist I have spoken to, or read about, who first learned about fission then: How could we possibly have been so stupid as not to have predicted it? The theory was there in front of their eyes. Bohr was just leaving for the United States, by ship, to spend the spring in Princeton. He was taking Rosenfeld. Bohr had a cabin in which a blackboard had been set up. He spent much of the voyage going over all the arguments for fission. He had made an arrangement with Frisch that he would say nothing until Frisch and Meitner published their paper. This took the pressure off of them to hurry. But Bohr forgot to tell Rosenfeld. As soon as they landed, Rosenfeld went off to Princeton and began to spread the word that very night. By the next day, the news had propagated all over the country and experiments were underway. The race for the atomic bomb was on.

In 1939, the Dutch physicist George Uhlenbeck emigrated to the United States. He had been a professor at the University of Utrecht. This left a vacancy at the university, and Rosenfeld was invited to fill it. Until that time he had been commuting between the University of Liège, where he had a position, and Copenhagen, where he was working with Bohr. Uhlenbeck left behind some unfinished business, including Pais. Pais had been studying at the University of Amsterdam and had decided that the physics being taught and practiced there was not up to date. He began to visit Uhlenbeck in Utrecht, and by 1938 he had enrolled at the university. When Uhlenbeck left, he introduced Pais to the very distinguished Dutch physicist Hendrik Kramers, who was at Leiden, and Kramers became Pais's mentor. Then Rosenfeld became Pais's official thesis advisor in 1940, when he took up his new post. In the meanwhile, the Germans had invaded Holland. Jews were banned from all civil service positions, including teaching at universities. Pais lost his assistantship. Rosenfeld arranged for him secretly to share the assistantship of his non-Jewish successor. In 1941, the Germans decreed that no Jew could be awarded a doctorate after June 14. Pais obtained his on June 9, the last Jew to be awarded such a degree until after the war.

Pais at first felt safe because of his former university status. But in 1943, the Germans began rounding up Jews from the universities intending to sent them east for extermination. Pais went into hiding. His sister and brother-in-law did not; they were captured, sent to concentration camps, and killed. Few people knew about Pais's hiding places. One of them was Kramers, who visited him and discussed physics. Pais's last hiding place was an apartment which he shared with his friend Lion Nordheim. They were betrayed and arrested by the Gestapo. Pais told me that he thought he was saved by his knowledge of German. He managed to persuade his captors that he was an innocent scholar. Nordheim was not so fortunate. He shared a cell with Pais, and Pais told me how he tried to comfort Nordheim the night before his execution.

Very soon thereafter, Holland fell to the Allies and Pais was liberated. Rosenfeld was Jewish only on his father's side.[8] Also, he had married a non-Jew. This, and the fact that he was a Belgian citizen, enabled him to remain in his post at the university until 1944. He then decided it was too dangerous and went into hiding in secret quarters in his attic. Although the Germans made a search, he remained undiscovered until Holland was liberated in 1945. Rosenfeld stayed in Holland until 1947, when he went to the University of Manchester to direct its physics department. He stayed there until 1958, when he became the first professor at the newly created Nordic Institute for Theoretical Atomic Physics (NORDITA) in Copenhagen. There he remained until his death in 1974.

Rosenfeld's wartime experiences seemed to have hardened his character. He is often described as a reserved, gentle, and even timid man. But after the war, when he began attacking people who were criticizing Bohr's interpretation of the quantum theory, he was ferocious. His letters and articles are scathing. To understand the issues, we must backtrack some years.

The colloquy between Bohr and Einstein concerning the quantum theory began in the late 1920s. Einstein presented various thought experiments which purported to show that the Heisenberg uncertainty principle led to contradictions. Bohr shot them down. This persuaded Einstein that there were no contradictions related to the uncertainty principle. Yet he still felt that the theory, despite its successes, did not give a complete description of reality. This was the purport of his paper with Podolsky and Rosen. But there is even a much simpler example than the one they gave in their paper. If one has a sample of radioactive elements, the theory does not tell you which of the

atoms will next decay. It gives you only the probability that one might decay. Moreover, there is no
way of using the classroom version of the quantum
theory even to formulate the proposition that a
particular atom decayed yesterday at four o'clock.
The quantum theory, as usually presented, deals
with the probability of future events and not a description of the past. If you think these issues de-
serve an explanation, then you might think that
Einstein had a case.

While this clash of Titans was going on, most
physicists were largely indifferent. There was too
much to do. The Nobelist I. I. Rabi once said to
me that he thought Bohr was very profound about
things that didn't matter. Rabi was trying to design experiments that would show a breakdown of
the theory. He never found any. After a visit to
Princeton in 1935, Robert Oppenheimer wrote to
his brother Frank, "Princeton is a madhouse: its
solipsistic luminaries shining in separate & help-
less desolation. Einstein is completely cuckoo."[9] Of
the founders, Heisenberg occasionally made philosophical comments—his version of Bohr's ideas.

Erwin Schrödinger, who was a little older than
the others and very philosophically minded, sided
with Einstein. He had been skeptical of the theory

for some time. In 1935, he published a paper in the German journal *Naturwissenschaften* whose translated title is "The Present Situation in Quantum Mechanics." In it, he introduced two things that have endured. One was a term of art: "entanglement." This describes the quantum state that is created, for example, in the Einstein, Podolsky, Rosen experiment. Even when the particles are widely separated in space—out of ordinary communication with each other—their properties are correlated. If you select one of these properties, the quantum theory predicts the probability, upon a measurement on one of these particles, of finding some possible value of this property. Once having found this, you can establish a correlation to the probable outcomes of measurements made on the distant particle. In this sense the particles are "entangled," despite the fact that no force is involved.

The second enduring thing Schrödinger introduced in that paper is the "cat." In this thought experiment, a cat is confined inside a sealed container. The container holds a radioactive sample. When it disintegrates it produces ionization, which unleashes a vial of poisonous gas, enough to kill the cat. At any given time, there is some probability of such a radioactive decay. In the meanwhile, if one wants to make a quantum mechanical description of the state of the cat, and wants to be provocative about it, one can say that at this time the cat is both alive and dead. A Bohrian positivist would probably say "prove it." You would then open the container to find either a live cat or a dead cat. I paid a visit to Schrödinger in his apartment in Vienna not long before his death in 1961. There were no cats. I was told that he did not like cats. Another founder, Dirac, so far as I know, never made any published comments about cats. He was too busy applying the theory. For example, the union of relativity and quantum mechanics that he created had, as an offspring, anti-matter. About this time, he noted, with his usual pith, that quantum mechanics explained most of physics and all of chemistry.

Bohr, however, soldiered on during the 1930s. He gave lectures and wrote papers. Out of this emerged something that is often referred to, though never by Bohr, as the "Copenhagen interpretation" of the quantum theory—the *Copenhagener Geist*. Since this was never spelled out formally, there are some differences of opinion as to what precisely it is. Yet all are agreed that Bohr's notion of "complementarity" plays an essential role. When I think of complementarity, what Saint Augustine said about time comes to mind. "What then is time? If no one asks me, I know what it is. If I wish to explain it to him who asks, I do not know." Substitute "complementarity" for "time." Bohr applied this idea, of which he was very proud, to practically all forms of human activity. I will confine myself to physics and give two examples, one classical and one quantum mechanical. I begin with the classical.

To this end, I consider a container which holds molecules of some gas. I then heat the walls of the container. This agitates the atoms out of which the walls are composed. When molecules of the gas collide with the walls, they are in turn agitated, and this agitation communicates itself to the entire gas. From a macroscopic point of view, we say that the gas has been "heated"; and if we insert a thermometer, it will show a rise in temperature. Temperature is a macroscopic property of the gas. If instead I choose to measure the energy of one of the molecules, I cannot in this measurement also measure the temperature of the gas, which is a collective property of the ensemble of molecules and requires a different sort of apparatus to measure. These two kind of measurements are, in Bohr's sense, "complementary." They apply to two mutually exclusive descriptions, both of which are valid in their domains of applicability. We must make a choice as to which one we wish to apply. Now to the quantum mechanical example.

In 1922, the French nobleman Louis de Broglie, who was working toward his doctorate, went to his thesis advisor, Paul Langevin, with an idea. Langevin found it so radical that he sent it to Einstein and asked his advice. Einstein said that he found it interesting and had been thinking along somewhat similar lines, although he had not published anything. De Broglie wrote a brief note which he published in the *Comptes rendus* of the French Academy of Sciences.[10] Reading this note is an odd experience. One knows where it led—but from the present point of view, it seems pretty confused. In more modern language, what de Broglie was suggesting, however obscurely, is that electrons—he deals specifically with electrons, which until that time had been assumed to be corpuscular—had a wave nature as well. He calls them "fictitious waves." He never really says what we would say now—namely, that the wavelength of these waves is inversely proportional to their momenta. Nor does he present any suggestion for an experimental test of this notion. Einstein did. Einstein said that if this was true, and if electrons were sent through a diffraction grating of a suitable size, diffraction patterns would emerge. This was the way Thomas Young and Augustin Fresnel, at the beginning of the nineteenth century, showed that light had wavelike properties. In 1927 Clinton Davisson and Lester Germer, and George Thomson independently, performed experiments that revealed the diffraction patterns of electrons. The "duality" of electrons was established. This was the counterpart to the duality of light—particles and waves—that Einstein had proposed in 1905.

The two aspects, particle and wave, were complementary in Bohr's sense. This manifests itself in expression for the wavelength in terms of the momentum. "Wavelength" is evidently a wave phenomenon, and momentum is something one associates with particles. Suppose we do an experiment with a diffraction grating that allows the electrons to pass one at a time. If we collect the electrons on a photographic plate, the regions of maximum and minimum intensity will spell out the diffraction pattern, but each electron will leave a localized spot on the plate—something that a wave does not do. Which aspect is revealed depends on our choice of experiment: complementarity. If one is not careful with language—and at times Bohr wasn't—this slips into idealism. Indeed, some writers, such as Heisenberg, did think that it was idealism—the notion of a "participatory universe." In the most extreme cases, the claim is that it is our presence that brings the laws of physics into being. It was precisely here that the conflicts with some of the Soviet scientists took place.

Vladimir Fock, who died in 1974, was a first-rate Russian theoretical physicist. He was not quite in the Nobel category, but some of the things he created are still taught. In 1935 he translated into Russian the Einstein, Podolsky, and Rosen paper. He agreed with Bohr's criticisms. Moreover, he saw complementarity as a manifestation of Engels' second dialectical law:[11] "The law of unity and the conflict of opposites." Fock also argued that Einstein's theory of relativity was a manifestation of "materialism." How sincere he was I do not know, but so long as Bohr did not stray into language which seemed to imply that matter did not have objective properties, Fock was all right. He seems to have come through the Great Purges of 1937 and 1938 unscathed. Others were not so fortunate.

Matvei Petrovich Bronstein was a physicist and astrophysicist of great promise. In the winter of

1938 he was put on a list of tens of thousands who were to be arrested, charged, and then executed. "Innocent until proven guilty" was considered a bourgeois notion. On February 18, 1938, he was charged and immediately executed. Whatever crime he had committed was never explained. His wife was told that he had been sentenced to ten years in labor camps without the right of correspondence. He was thirty-two. Lev Davidovich Landau was, in most people's opinion, the greatest Russian theoretical physicist of the twentieth century. He was awarded the Nobel Prize in 1962. In 1938, he was arrested on a trumped-up charge and imprisoned by the NKVD (the Soviet police). If a very courageous colleague, Pyotr Kapitsa, had not written a letter directly to Stalin, Landau might never have gotten out. Unlike Fock, both of these men were Jewish, which made everything worse for them. In June 1941, the Germans invaded the Soviet Union, and the compatibility of dialectical materialism and complementarity was not high on the list of priorities.

I began my graduate studies in the early 1950s. English translations of some excellent Russian texts were available. I remember in particular a book entitled *Quantum Electrodynamics*, written by A. I.

Akhiezer and V. B. Berestetski; the translation appeared in 1953. You would be reading along in the technical material when suddenly, and with no warning, there would be a paragraph informing you how this work was connected to dialectical materialism. Then the physics went on as if nothing had happened. I always thought of these as little commercial messages from the "sponsor." It did not occur to me to inquire what circumstances must have led to this. In fact, after the war, the ideological struggle for the soul of the quantum theory in the Soviet Union took a rather dramatic turn. In the 1930s the views of Fock were criticized, but those criticisms could be largely ignored. At least, Fock more or less ignored them. But by the late 1940s things had changed. Fock now had critics he could not ignore. Among them was the physicist D. I. Blokhintsev.

Blokhintsev, who was a few years younger than Fock, was a theoretical physicist of considerable distinction. He was, one gathers, a bit higher in the ranks of the Establishment. He ended his career as chairman of the department of theoretical nuclear physics at the Moscow State University. In 1944, he published the first edition of a text on the quantum theory. It went though several editions and translations. In the first edition, he embraced at least part of the Copenhagen interpretation.[12] To understand this, and Blokhintsev's later changes, we have to back up a little.

De Broglie was very vague about the meaning of his waves. He did not provide an equation that described their propagation in space and time. This was done in 1926, in a series of four magisterial papers by Erwin Schrödinger. The Schrödinger wave equation tells us how the "wave function," usually denoted by ψ, evolves in space and time. But what does it mean? At first, Schrödinger thought that it was a guide wave that guided the particles. This guide wave was to have the same physical properties as any wave. But quite soon this interpretation was shown to be untenable. In 1927, the Göttingen theoretical physicist Max Born proposed that these waves were waves of probability. A particle was likely to be found where the amplitude of the wave was large. This is the interpretation we still have.

The question then arose: Is this probability all there is? To this, Bohr answered yes. And Fock agreed. In his introduction to his translation of the Einstein, Podolsky, and Rosen paper, Fock wrote: "In quantum mechanics the conception of state is

merged with the conception of 'information about
the state obtained as a result of a specifically maximally accurate operation' [a measurement of, say,
the position of a particle]. In quantum mechanics the wave function describes not the state in
the usual sense, but rather 'information about the
state.'"[13]

We are in the position of someone at a horse race
who is allowed to know only the odds. No other
information, such as the breeding of the horses, is
available. If we go further and say that no such information is ever available to anyone, we get some
sense of the quantum mechanical situation.

In the 1944 edition of his book, Blokhintsev did
not seem to disagree with this, but in the 1949 edition he wrote: "The chapter which concerns concepts of the state in quantum mechanics has been
changed, and the clarity of the discussion of the
uncertainty relationships has been improved. In
the new edition of the book, ideological questions
connected with quantum mechanics are also considered, and the idealistic conceptions of quantum
mechanics which are now widespread abroad are
subjected to criticisms."[14]

In the interval between the editions, various Soviet authors had offered some strong and poten-

tially dangerous criticisms of the Copenhagen interpretation. In the 1949 edition, Blokhintsev speaks of the wave function controlling the destiny of what he called "microparticles," and these had an objective reality. He and Fock had some spirited exchanges which did not seem to change either of their minds. In 1966, Fock summarized his position:

> Dialectics plays an essential role in obtaining new outlooks on the external world and the appropriate ways of its description. To solve the contradictions between classical and quantum mechanical description, between causal and probabilistic laws, between wave properties and corpuscular properties of matter, between elementarity and mutual transformability of particles, between objectivity of micro-objects and the necessity of introducing measuring instruments to describe their properties, etc., etc.—to solve any of these problems, a dialectical approach is quite indispensable. Before proceeding to the solution, one must first of all realize that the contradictions are of a dialectical nature, so that they must vanish if a more general and more appropriate formulation of laws of nature is found. (In the early history of quantum mechanics, an instance of that proceeding is given by the conscious application of dialectics by Bohr, who formulated the "semi-classical" quan-

tum mechanical laws in spite of the apparent contradictions later solved by the [wave] quantum
mechanics. . . . Dialectical materialism is a living
and not a dogmatic philosophy. It helps to give the
experience obtained in one of the domains of science a formulation of such generality that it may
be applied to other domains.[15]

In contrast, in the 1964 edition of Blokhintsev,
which I have, there is not a single mention of dialectical materialism.

Rosenfeld observed all of this with a good deal
of contempt. So far as he was concerned, the quantum theory did not have "interpretations." There is
no quantum theory but the quantum theory, and
Bohr was its messenger. In 1972, the American
physicist Henry Stapp published an article entitled, "The Copenhagen Interpretation."[16] He sent a
draft to Rosenfeld for vetting and got back this response:

I notice from your further letters with new title pages that you hesitate about the best title for
your essay. I have no very strong views about this,
but I would incline to prefer your March 31 title
["Quantum Theory, Pragmatism and the Nature
of Space-Time"], the reason being that it does not
contain the phrase "Copenhagen interpretation,"
which we in Copenhagen do not like at all. Indeed,

this expression was invented, and is used by people wishing to suggest that there may be other interpretations of the Schrödinger equation, namely their own muddled ones. Moreover, as you yourself point out, the same people apply this designation to the wildest misrepresentations of the situation. Perhaps a way out of this difficulty would be for you to say, after having pointed out what the difficulty is, that you make use of the phrase "Copenhagen interpretation" in the uniquely defined sense in which it is understood by all physicists who make a correct use of quantum mechanics.[17]

Rosenfeld was at his most vitriolic when he addressed fellow Marxists. Typical of the genre is a letter he sent to Frédéric Joliot-Curie in 1952. Joliot, who had shared the 1935 Nobel Prize in Chemistry with his wife, Irène, had fought in the Resistance and was a member of the French Communist Party. Rosenfeld writes,

I think it is my duty to inform you of a situation which I consider quite serious. It concerns your "foals" [Jean-Pierre] Vigier, [Evry] Schatzman, [Gérard] Vassails, and the whole lot, all young intelligent people, and all full of desire to do well. Unfortunately, at the moment they are quite sick. They have gotten it into their heads that it is necessary persistently to shoot down complementarity and save determinism. The illfortune is that they have not understood the problem and—what is even worse—they have made no serious effort to understand it. I have done what I could to redeem them. . . . I have taken pains to do an explicit Marxist analysis of the question and clearly show the simultaneously dialectical and materialistic character of complementarity. As the only response, Schatzman sent me a polemical writing full of incorrect physics and quotations from Stalin, which, he maintains, through a casuistry that frightens me, oppose physical evidence. This reveals a profound crisis among these young people, and it is high time (if it is not already too late) to straighten them out. They are under a spell of a scholasticism which borrows the external forms of Marxism, but is as opposed to its genuine spirit as is the blackest Catholicism. The best Soviet physicists are subject to attacks from this scholasticism, which creates even more havoc in Moscow than in Paris. Surely it would be desirable that the French physicists show themselves capable of distinguishing the wheat from the chaff. At the moment, these young fanatics are the laughingstock of the theoreticians such as [Jean-Louis] Destouches and de Broglie. I am ready to give you all possible support to redress the situation on the basis of the ideas explained in my paper. But it is up to you, my dear Joliot, to take the initiative.[18]

While the "foals" might be brought into line with a good talking to, David Bohm was another matter. His apostasy was irredeemable. He was a *soi-disant* Marxist who had abandoned the laws set down by Marx and Engels. Still worse, he had produced a version of quantum mechanics which, if adopted, would render Rosenfeld's lifelong work irrelevant. In Bohmian mechanics, there is no complementarity. There is no Einstein, Podolsky, Rosen paradox. Quantum mechanical measurements are just another set of ordinary interactions. The theory is deterministic and essentially classical. Classical particles are guided by waves determined by a Schrödinger equation. In the usual case, the solutions to the Schrödinger equation predict the probabilities of the outcomes of experiments. Here, the solutions provide the information that is needed to determine the trajectories of classical particles. For Rosenfeld, Bohm and his mechanics were beyond the pale.

Bohm's two articles in the *Physical Review* appeared in January 1952. By this time, Bohm was in Brazil. He had been there since the previous November.[19] He had appeared before the House Un-American Activities Committee, taken the Fifth Amendment, been indicted by a grand jury for contempt of Congress, and finally acquitted on May 31, 1951. Bohm had joined the Communist Party in November of 1942, but remained a mem-

ber only for some nine months. But he defended
the Soviet regime until Stalin's death. After Bohm's
appearance before the House committee, in which
he took the Fifth Amendment, he received written
notice that he was no longer to have access to the
Princeton campus. How this was to have been en-
forced is unclear. In fact, the department chairman, Allen Shenstone, told Bohm that he would be
welcome in his office at any time, and Bohm continued working with his graduate students.[20] Princeton University had put him on paid leave, and
then refused to renew his appointment in June
1951. Some Brazilian physicists who were in Princeton were able to find him a job at the University of
São Paulo. Not long after Bohm arrived in Brazil,
he was asked to present his passport to the American consul's office for "verification." It was confis-
cated, and he was told that it would not be returned
unless he returned to the United States to reclaim
it. Fearing that he might be subject to further legal
action if he returned, Bohm took Brazilian citizenship. It was some thirty years before his American
citizenship was restored. By this time, he had become an intellectual celebrity.

Bohm's three-year stay in Brazil was not an especially happy one. He found the country techno-

logically backward and somewhat "dirty." He also managed to get into a fight with some of the physics professors. Bohm was given a stipend he could use to hire assistants. He chose them from the United States, and this was deeply resented. Two of them, Ralph Schiller and George Yevick, happened to have been colleagues of mine at the Stevens Institute of Technology. In the three decades that I knew them, neither one ever said a word to me about this experience. In any event, there was a nasty fight over such appointments. At one point the German physicist C. F. von Weizsäcker appeared in Brazil and seemed to side with Bohm's opponents, whom Bohm referred to as "Nazis." It must have been a relief all around when Bohm left for Israel, where he stayed for two years before finally settling in Great Britain, at Birkbeck College in London. Here he remained until his death in 1992.

In the midst of Bohm's turbulent activities in Brazil, Rosenfeld came to give lectures on the foundations of the quantum theory. Bohm attended the lectures. Afterward he wrote a letter in which he said, "Prof. Rosenfeld visited Brazil recently, and we had a rather hot and extended discussion in São

Paulo, following a seminar he gave on the foundations of the quantum theory. However, I think we both learned something from the seminar. Rosenfeld admitted to me afterward that he could at least see that my point of view was a possible one, although he personally did not like it."[21]

Bohm must have read into Rosenfeld's reaction what he wanted to see. There was no way Rosenfeld could have thought that Bohm's version of the quantum theory was a "possible" one. Sometime after his visit to Brazil, Rosenfeld wrote a letter to the British philosopher of science Lancelot Law Whyte in which he said,

> Why then . . . do I bother to write all these articles and reviews? I am not concerned with Bohm's own salvation: he is past healing. . . . All I actually wish to do is to sound a warning to the bewildered students, who don't know what to make of it, and to interested outsiders, who— however clear-minded and critical—cannot reasonably be expected to be sufficiently familiar with the technicalities of the case to discover themselves the utter emptiness of the claims of Bohmism. I realize that I may give the impression of being "intolerant" and "dogmatic" in spite of the fact that I have been all these years advocating

tolerance and free thinking . . . and waging a fight against the modern brand of dogmatism which represents the greatest threat to our scientific tradition, viz pseudo-marxist theology.[22]

At present, this discussion of Marxism and science seems almost antique. One has the impression that Russian science has now moved on. I at least have not found the sort of ideological insertions in Russian papers that used to occur. But when I think of the past, I am reminded of an anecdote I heard. In 1944, Lavrenty Beria, the chief of the Soviet secret police, was appointed by Stalin to direct the nascent nuclear weapons program, which involved some applied quantum mechanics. Beria discovered that certain of the physicists were straying off the ideological reservation, and he complained to Stalin. Stalin allegedly said to him, "Leave my physicists alone. We can always shoot them later."

:: 4 ::

A Double Slit

We chose to examine a phenomenon which is impossible, absolutely impossible, to explain in any
classical way, and which has in it the heart of
quantum mechanics. In reality it contains the only
mystery.

—Richard Feynman[1]

The particle world is the dream world of the intelligence officer. An electron can be here or there at
the same moment. You can choose. It can go from
here to there without going in between; it can pass
through two doors at the same time, or from one
door to another by a path which is there for all to
see, until someone looks, and then the act of looking has made it take a different path. Its move-
ments cannot be anticipated, because it has no
reasons. It defeats surveillance because when you
know what it is doing you can't be certain where
it is, and when you know where it is you can't
be certain what it's doing: Heisenberg's uncertainty principle; and this is not because you're not
looking carefully enough, it is because there is no
such thing as an electron with a definite position

and definite momentum; you fix one, you lose the
other, and it's all done without tricks, it's the real
world, it is awake.

—Tom Stoppard, *Hapgood*[2]

IN THE FALL OF 1947, I entered Harvard College as a seventeen-year-old freshman. I had no
real ideas about my future except that I was quite
sure that it would have nothing to do with science.
I had no interest in science. But at that time the
president of the university was James Bryant Co-
nant, a chemist who had played a very important
role in the creation of nuclear weapons. The experience had left him with a good deal of concern
about this new force in the world. One manifestation of his concern was the notion that every Harvard graduate should have some sort of scientific
education. He knew, of course, that most of us
would not have careers in science, but we might
well have careers where we would have to make
decisions that involved science. So under the rubric
of "General Education," a set of natural-science
courses had been instituted which presented non-
mathematical surveys of different scientific fields.
Every Harvard graduate was required to pass only
one of them, unless he (there were no she's) was a

science major. You also had to be able to swim two laps of a twenty-five-yard-long pool. Faced with the former obstacle, I consulted the Harvard Confidential Guide to Courses—a small booklet produced and sold by the undergraduate newspaper, the *Harvard Crimson*—to find out which course was considered the easiest. The vote among the undergraduates who took the various courses and compared notes was unanimous: it was Natural Sciences 3, taught by the historian of science I. Bernard Cohen.

In the interests of full disclosure, I must admit that I had had a physics course in high school. I have searched my memory to recall anything I learned in that course and have come up blank. It was my only contact with physics in high school. I also took the required mathematics courses, in which I was a good student. If someone had told me that there were professional physicists and mathematicians who did this sort of thing for a living, I would not have believed it. I had never met such a person. I had of course heard of Einstein, but I had no idea of what he actually did. I also had no scientific curiosity. I tell you these things to explain what kind of student I was when I enrolled in Natural Sciences 3.

Cohen was a fine lecturer for this level of student. There were well over a hundred of us. He had a rich, deep voice and a round, reassuring handwriting when he wrote on the blackboards. We began with the Greeks, worked our way along to Copernicus, Galileo, and then Newton. Cohen was a Newton scholar so we learned about Newton's life. For the first time I realized what a very strange man he had been. His masterwork, the *Principia*, which he wrote at the end of the seventeenth century, was deliberately presented in such a way that it would be accessible only to scholars. It was written in Latin and used geometric arguments that could have been replaced by much simpler demonstrations using the calculus—which Newton had invented. Before it was published, Newton got into a nasty priority fight with Robert Hooke as to which one of them had first enunciated the correct law of gravitation. They had previously clashed over their work on light. On the other hand, Newton's *Opticks*, which was published in 1704, is a very different kind of book.

In the first place, it is written in English and the style is as congenial as Newton was capable of. It deals in large measure with the optical experiments he performed, including the one with the prism that showed that white light was really a mixture of light of many colors. Unlike the *Principia*, where Newton avoided speculation about the ultimate nature of things, in the *Opticks* he does speculate. There is the famous Query 31, in which he talks about the atomic theory.

All these things being considered, it seems probable to me, that God, in the Beginning, form'd, Matter in solid, massy, hard, impenetrable Particles, of such Sizes and Figures, and with such other Properties and in such Proportion to Space, as most conduced to the End for which he form'd them, and that these primitive Particles being Solids, are incomparably harder than any porous Bodies compounded of them; even so hard, as never to wear or break in pieces; no ordinary Power being able to divide what God himself made one in the first Creation. While the Particles continue entire, they may compose Bodies of one and the same Nature and Texture in all Ages. But should they wear away, or break in pieces, the Nature of Things depending on them, would be changed. Water and Earth, composed of old worn Particles and Fragments of Particles, would not be of the same Nature and Texture now, with Water and Earth composed of entire Particles in the Beginning. And therefore, that Nature may be lasting, the Changes of corporeal Things are to be placed only in the various separations and new Associations and Motions of these permanent Particles, compound Bodies being apt to break, not in the midst of solid Particles, but where these Particles are laid together, and only touch in a few points.

It is clear that when it came to material objects, Newton was an atomist *pur et dur*. But what about light? Here it seems to me that he was somewhat more nuanced. He writes, "In the first Books of these Opticks, I proceeded by the Analysis to discover and prove the original Differences of the Rays of Light in respect of Refrangibility, Reflexibility, and Colour, and their alternate Fits of easy Reflexion and easy Transmission, and the properties of Bodies, both opaque and pellucid on which their Reflexions and Colours depend." I am not sure what meaning is to be attached to "alternate Fits of easy Reflexion and easy Transmission," but it does not sound like a simple particle picture to me, although this is how Newton was later interpreted. Opposed to this view was, in the first instance, that of Hooke, who claimed that light was a wave. But of greater significance was the work of Christian Huygens (or Hugens), the great Dutch contemporary of Newton, whom he came to visit in

1689, before Newton's *Opticks* had been published. I do not know if they discussed the nature of light. Huygens own *Treatise on Light* was published in French in 1690. He proposed that light was an oscillation of waves in an aetherial medium. What he did not do was to consider the interference of waves. When two waves meet, they can reinforce each other or interfere with each other destructively. Huygens' "waves" were actually pulses, and not waves as we would think of them.[3] Nonetheless, either concept of light could account for the observations that were then available. Everything changed with the work of the early nineteenth-century British polymath Thomas Young. It is to him that we owe the idea that light waves can interfere with each other.

Over the years I have encountered a certain number of people whom I would label geniuses. I had an afternoon tea with Schrödinger in his apartment in Vienna, and had a chance to show Dirac around that city. When I was in high school I spent some time with Duke Ellington. As I've discussed, I had one memorable lunch with W. H. Auden. And when I was a Harvard undergraduate, I had the opportunity to ask John von Neumann a question. The question was whether he thought the computing machine would ever replace the human mathematician. He response was, "Sonny, don't worry about it." What these people had in common was their ability to do things with apparent ease that most of us cannot do at all. By this standard, or any other, Thomas Young was a genius.

He was born into a Quaker family in Milverton in Somerset, England, in 1773. By the age of fourteen he not only knew Latin and Greek, but had a grounding in French, Italian, Hebrew, Chaldean, Syriac, Samaritan, Persian, Turkish, and Amharic. Later in life he made basic contributions to the decipherment of Egyptian hieroglyphics, something that was completed by the French linguist Jean-François Champollion in 1822. Young, who became independently wealthy, trained as a physician in London and made significant contributions to the practice of medicine. This is not what made him immortal to physicists. Prior to the turn of the century, he began conducting experiments in physics. The ones that he made on light were summarized in his November 1801 Bakerian Lecture— "On the Theory of Light and Colours"—before the Royal Society, of which he had been a fellow since 1794. It was these experiments that persuaded most physicists that Huygens was right: light is a wave.

The experiment that he described in his Bakerian Lecture is not the one that is usually explained to students. That came later. In this one, which was actually first done by Newton, he made a square hole in a card. The breadth of the hole, he notes, was "66/1000" of an inch. Over the hole he attached a hair, which stretched across it from top to bottom. It had a diameter of 1/600 of an inch. Next he shined candlelight on this arrangement. Nowadays we would use lasers, which produce a concentrated beam. The hair split the light beam into two parts. I think of those bicycle races where the stream of riders is split in two by a traffic island in the middle of the road. After they pass the island, the two streams reunite and to all appearances look as they did before. This is what the particle theory of light would predict. But this is not what Young observed. On the other side of the hair, light and dark fringes were observed. Young understood immediately the implications of this. It had to do with the interference of light waves. Later he repeated this experiment with two pinholes instead of the hair, and got similar results. This was the Young "double-slit" experiment. While all of this was interesting to me as Cohen explained it, it was hardly life changing. What happened next was.

Near the end of the first semester, Cohen presented a couple of lectures on the theory of relativity. I was overwhelmed. The idea that the speed of light was the universal speed limit and that every observer in a state of uniform motion would observe the same speed was incredible to me. There was also the claim that clocks in motion appear to run slower when observed by someone at rest. This also seemed incredible to me. What was the man talking about? In the course of this he said—I think as a joke, although I did not see it that way—that only ten, or maybe it was twelve, people in the world understood the theory. Much later I learned that when the great British astronomer Arthur Eddington, who in the 1920s had written a noted monograph on relativity, was asked if only three people in the world understood it, he replied, "Who is the third?" Eddington was certainly referring to Einstein's general theory of relativity and gravitation, which at the time was not widely understood. In any event, with all the lunar audacity of a Harvard freshman, I decided to become the eleventh or thirteenth person to understand the theory. I had no idea that there was a special theory and a general theory. I had a plan: I would find a book that explained it.

Not only did I not have the remotest background, mathematical or otherwise, to understand the theory, I did not even know what it meant to understand something like relativity. In high school I had taken Spanish, so I knew what was meant by understanding a passage in Spanish. You translated it into English, which presumably you "understood." If you came to a hard word, you looked it up in a dictionary, where it was "defined" in terms of words that you "understood." When it came to poetry, what was meant by "understanding" was to be able to describe its meaning in terms of prose. In geometry, what was meant by "understanding" was the capacity to prove something from the axioms, perhaps with the aid of diagrams. But what does it mean to understand the theory of relativity, or the quantum theory for that matter? What it does not mean is to be able to "explain" the theory in terms of the theory that preceded it. If anything, this process should be the reverse. Be all that as it may, I went off to Widener Library to find a book.

I thought that I might as well get a book by Einstein, since he surely understood the theory. There were a couple, and I picked one called *The Meaning of Relativity*. I liked the title. It was the worst choice I could possibly have made. It was based on a series of lectures that Einstein delivered to specialists in 1921 at Princeton, although it has had revised editions. It is very sophisticated and highly mathematical. Apart from the opening sentence, "The theory of relativity is intimately connected with the theory of space and time," I understood essentially nothing.[4] I might have given up, but instead I went to Cohen. He might have dismissed me as just another foolish freshman, but he didn't. He told me that in the spring semester there would be a course at about the same level as his which would be devoted to modern physics. He told me that it would be taught by a man named Philipp Frank, who was actually a friend of Einstein's and had just written a biography of him. Cohen said he would have no objections if I took that course along with his, so I signed up.

The course met once a week, I think on Wednesday afternoons, for two hours. It met in the large lecture hall in the Jefferson physics laboratories. When I showed up for the first lecture, the room was full. Later I learned that people from other universities in the area came to audit. I had no idea what to expect. What would Professor Frank look like? When I first saw him, he seemed perfect for the part. He was a smallish man who walked with a bit of a limp. What hair he had left was in kind of a halo around his head. His accent was not easy to place. Although he had been born in Vienna (in 1884), he had lived in Prague from 1912, when he succeeded Einstein at the German University, remaining until 1938. Then he came to the United States, where Harvard cobbled together some kind of appointment for him. He knew all sorts of languages, including Persian, which he had studied in night school in Vienna. I envisioned his languages as being piled up on top of each other like the buried cities of Troy. When he spoke English, the shards of the other languages appeared. Professor Frank would lecture for an hour and then make what he called "a certain interval." After the "interval," he would answer questions or go into things with more depth, "if you know a little of mathematics."

His course began with the Greeks. We learned that the Greek astronomers were committed to the notion that the celestial objects were attached to uniformly rotating spheres. But this did not fit the facts. Planets, for example, periodically appear to move backward. To account for this and still to "save the appearances," astronomers had the uniformly moving planets moving in spheres around spheres—epicycles. Cohen had discussed this, but Professor Frank added a new insight. Kepler had replaced all of this with single elliptical orbits. Professor Frank noted that such a motion can always be reproduced as a series of uniform circular motions, so the two models cannot be distinguished if all one is interested in is describing the observed motions. Other criteria enter, such as simplicity or "elegance." When it came to Newton, Professor Frank explained the story of the falling apple. He said imagine an apple on a tree that grows to be so tall that it extends from here to the Moon. Then the Moon is an apple on the tree, and it too must fall in the field of the Earth's gravity. Then we came to relativity.

Professor Frank explained the relativity principle which was first stated by Galileo: the laws of physics are the same for observers at rest and observers moving uniformly with respect to these observers. Indeed, the moving observers can perfectly well claim that they are the ones who are at rest. He told us that Einstein from very early on believed that this must apply to all laws of physics but realized there was a contradiction. If he was allowed to move at the speed of light, he could ride on a light wave. Then it would no longer look like a wave; so he would know that he was moving at the speed of light, which violates the relativity principle. It took him many years from his having first thought of this paradox to resolving it. You cannot in relativity move at the speed of light.

Professor Frank had a simple way of explaining the transformations that take you from one system of coordinates to another that is moving uniformly. I have used it often when teaching the subject. He also told a funny story. He'd visited Einstein in Berlin a few years after Einstein had published his popular book on relativity. By this time Einstein had married his second wife and was living with his two step-daughters, one of whom was present on an occasion when Einstein and Frank were conversing. Einstein said his book was so simple that even his step-daughter could understand it. He then left the room and Professor Frank asked her if it was really true that she understood it. Yes, she said, she understood everything except what a coordinate system was.

Following relativity, Professor Frank turned to

light. He told us about Young's double-slit experiment and noted that if Young had had more modern equipment he would have seen wavelike manifestations even from a single slit. There would be a buildup of intensities in front of the slit, but then there would be lesser peaks at spaces to either side. The profile would look like that of a mountain with its shorter companion peaks to either side. All of this showed that in these experiments light behaved like a wave. But then Professor Frank told us about one of Einstein's 1905 papers—actually, the one for which he won the Nobel Prize—in which he argued that in other experimental arrangements light would behave like a particle. In particular, if you shined light on a metallic surface, electrons would be liberated from the metal: the photoelectric effect. This could also be predicted from classical physics. But what Einstein was proposing was that the energy of these liberated electrons depended not on the intensity of the incident light, but only on its color. The more violet the light, the more energetic the liberated electrons. In these experiments, light behaved like a stream of particles—quanta—whose energy was proportional to the frequency of the radiation. Professor Frank said that if the intensity of the light was diminished so that it consisted of a single quantum, this quantum could still supply the energy to liberate an electron. What is light? It is a wave or a particle? It is both and neither. Then we turned to electrons.

Professor Frank told us about the proposal of Louis de Broglie: that under certain conditions electrons exhibited a wave character. He told us about the experiments—analogous to those of Young—in which electrons were made to impinge on a grating of slits. They were then deposited on a detector, and one could observe the same kind of lines of intensity that Young had observed for light. What were electrons—waves, particles, both, neither? Electrons, Professor Frank said, are electrons. But then he suggested a thought experiment—one that he said had never been carried out. It has now.[5] Suppose you made a double slit and your beam of electrons was so weak that only one electron at a time went through the system. What would happen? If both slits were open, Professor Frank told us, then over the course of time—a least according to the quantum theory—the stream of single electrons would re-create the whole pattern. The pattern would not be, as one would expect for particles, the sum of the two open-slit patterns. It would show the whole effect of the interference. There was no way of predicting where a given electron would land; but using the theory, you could predict where it was most likely to land.

This, to me, was already amazing, but what he said next was more than amazing. It ultimately convinced me to go into physics. Suppose you closed one of the slits and allowed the electrons to enter the apparatus one after the other. What would happen? The two-slit pattern would disappear and be replaced over the course of time by the one-slit pattern. The question that immediately occurred to me was how does the single electron going though the slit "know" that the other slit has been closed? Where does that information come from? Today I know that quantum mechanics does not provide an answer. In fact, it is not even the right question to ask. But at the time, I didn't know this and was completely baffled and became determined to learn more. And so I began a long course of study, eventually ending up with a Ph.D. in physics.[6]

I have often wondered if young people nowadays, when they are first exposed to the quantum theory, have the same sense of wonderment that I did. After all, when I first learned it, the theory was only about twenty years old. All of the creators were still around lecturing and writing. Now the theory is well over a half-century old. Do young people simply accept the rules and get on with it? What I do know is that some nonphysicists do have this sense of wonder—even if, or mainly if, they do not really understand the theory. But they have absorbed enough so that it has entered their creative imagination. A very interesting example is the playwright Tom Stoppard.

In the interests of full disclosure once again, I have to confess that I am a Stoppardian of the deepest dye. Plays like *Jumpers* or *Travesties* are for me intellectual champagne. The mixture of ideas and wordplay are for me irresistible. In the fall of 1994 I received a phone call from a woman named Anne Cattaneo. She identified herself as the editor of a journal named the *New Theater Review*—now called the *Lincoln Center Theater Review*. She told me that Lincoln Center was mounting a play by Stoppard called *Hapgood*, which had premiered in London in 1988. She also told me that the play involved quantum mechanics in some way. I had read about such a play and had been intrigued. Stoppard and quantum mechanics— what a pairing. Then she went on to tell me that the *New Theater Review* was planning an issue devoted to the play and its scientific implications. Would I like to contribute? She said that Stoppard was going to be one of the contributors. The prospect of sharing the stage with Stoppard was irresistible, so I agreed. The first thing I did was to buy a copy of the play.

Hapgood is both the name of the play and the name of the principal character. Her first name, which is almost never used in the play, is Elizabeth —Lilya in Russian. She is a spy, indeed the head of a coven of spies who refer to her as "mother." They are British except for one, Joseph Kerner, who is Russian. It was Hapgood who turned him, but in the course of things the turning got a bit out of hand. Hapgood became pregnant and gave birth to a son, Joe. He is an adolescent and does not know who his father is. He is much more interested in playing rugby. Some of the play takes place during his school games. Kerner works at CERN, the elementary-particle laboratory in Geneva. He is working on an anti-ballistic missile warhead that involves anti-matter. This is only marginally more absurd than other anti-ballistic missile programs I have heard about. There is a convoluted plot involving Kerner's giving some or all of this information to the Russians. Periodically, Kerner gives brief lectures about the quantum theory and its relevance to human behavior. Schrödinger had enough trouble applying quantum theory to cats, let alone human beings. I will shortly deconstruct one of these lectures, but first let me say something about how Stoppard appears to have gotten into this. What I know is what he wrote in the *New Theater Review*—an article which he entitled "The Matter of Metaphor." My article, which followed his, was called "A Trick of Light" and covered some of the same ground I have been dealing with here, though in considerably less detail.

This is Stoppard: "Science for non-scientists is a boom area in publishing, and *Hapgood* is itself a fruit of a dozen books about quantum physics written for a general readership. If there were ever a general reader, I am one. I am not even a closet scientist, and certainly not a frustrated scientist. In fact, I probably owe it to my general lack of scientific education that the central image of *Hapgood*—the dual nature of light—excited me so much when I finally caught up with what is wearily familiar to anyone who stuck with physics past high school."[7] "Wearily familiar" does not describe my experience, as I have tried to explain.

Stoppard goes on: "But, of course, it excited me for its potential as a metaphor. *Hapgood* is not 'about' physics, it's is about dualities. No—let the playwright correct the critic in me—Hapgood is not about dualities, of course, it's about a woman called Hapgood and what happened to her between Wednesday morning and Saturday afternoon 1989 just before the Berlin Wall was breached."

Stoppard continues, "As a matter of fact, the story has much more to do with espionage than physics, but I won't deflect any compliments that might be going for a play with a reasonably plausible physicist on board, because the springs of the play are indeed science; it was only in looking around for a real-world metaphor, that I hit upon the le Carré world of agents and double agents. The physics came first, the woman called Hapgood came second."

And he adds, "Incidentally, while several of the physicists who saw the play went out of their way to be kind about its general veracity, John le Carré, whom I came to know and, I seem to recall, who went on my tickets, managed to avoid mentioning the play from that moment on; which I immediately recognized as the height of courtesy."[8]

Stoppard does not tell us which physicists vouched for the play's "general veracity," or

whether read it as well as saw it. Things go by
pretty quickly in an actual performance. Nor does
he tell us which popular books he read. It would be
interesting to know. We do know that he looked at
Feynman's lectures, since he quotes from them in
the prelude to the printed edition of *Hapgood*. This
is a bit like my consulting Einstein's Princeton lectures to learn about relativity. Feynman gave these
lectures to undergraduates at the California Institute of Technology (Caltech).Feynman being Feynman, they are full of original insights. They are
highly technical. I was told that the undergraduates found them pretty tough going, although the
faculty members, who audited them in droves,
loved them. It is the last place I would send a non-
scientist who wanted to get a flavor of the quantum
theory. I was also told that Stoppard visited Caltech
and actually sat for a while in Feynman's empty
office—osmosis.

What I would like to do now is to deconstruct
one of Kerner's mini-lectures on the quantum theory. I do this in the spirit of the seminars that Bohr
used to moderate where his many interruptions of
the speaker were always accompanied by his saying that he was not there to criticize, but only to
learn.

Kerner is talking to Paul Blair, another agent,

who has informed Kerner that his cover as a double agent has been "blown." Kerner has reached the end of his career, "except as a scientist." Kerner makes the point that a double agent is not like a giraffe—something definite but "more like a trick of the light." Then he explains, "Look. [He points.] Look at the edge of the shadow. It is straight like the edge of the wall that makes it. This means that light is particles: little bullets. Bullets go straight. They cannot bend round the wall a little, like the water round in a stone in the river."

I am surprised that none of the physicists Stoppard had contact with informed him that this characterization of light is entirely wrong. It had been observed as early as the seventeenth century that shadows cast by light were fuzzy. Hooke and Huygens argued that this implied that light had a wave nature, which manifested itself in this kind of observation. Light waves can interfere with each other, whereas particles, as we usually conceive of them, cannot. When Einstein wrote his 1905 paper, which claimed that under some circumstances light could exhibit particle behavior, he began it by explaining how such a thing was possible, given all the evidence that it was a wave. Thus began the idea of the complementary nature of light.

The dialogue continues:

Blair: [Irritated] Yes. Absolutely.

Kerner: So that's what. When you shine light through a gap in the wall it's particles. Unfortunately, when you shine light through *two* little gaps, side by side, you don't get particle pattern like for bullets, you get wave pattern like for water. The two beams of light mix together and . . ."

I have already explained that if you do the one-slit experiment with light, or with electrons for that matter, you get a light-and-dark pattern on the detecting screen. This happens because different parts of the slit produce waves that interfere with one another. Even in the one-slit case, the sort of question that bothered me as a freshman arises. Suppose you let the electrons through one at a time. Each electron will land at some spot on the detector, but as they accumulate they collectively reproduce the pattern of light and dark lines. How does the single electron know how to do this? How does it know where to go to play its role in the final pattern?

Conventional quantum mechanics has an answer, but it is not one that would have satisfied me as a freshman. By now I am used to it. The equation devised by Schrödinger has a solution which is a function of space and time. It you take this solution and square it, you get the probability of finding the electron, say, at some place at a given time.[9] Where the function is large the probability of finding the electron is greatest. Nothing in the theory tells you where a given electron will land; you find out only where it is *likely* to land. This is one of the things that troubled Einstein. It also troubled David Bohm, who produced his alternate interpretation of the quantum theory, with its guide waves. The answer to the question of how an electron knows what to do in Bohmian mechanics is that the guide wave tells it.

Back to the play. Blair is asking Kerner which side he's on:

Blair: Joseph, I want to know if you're ours or theirs, that's all.

Kerner: I'm telling you but you're not listening. Now we come to the exciting part. We will watch the bullets to see how they make waves. This is not difficult, the apparatus is simple. So we look carefully, and we see the bullets, one at a time. Some go through one gap and some go through the other gap. No problem. Now we come to my favourite bit. The wave packet has disappeared. It has become particle again.

In 1927 there was a conference of leading physicists in Como, Italy, to celebrate the centenary of the death of Alessandro Volta. Quantum mechanics was something like two years old. Heisenberg had just published his uncertainty principle. Both Bohr and Einstein were at the conference. Bohr introduced his notion of complementarity. Einstein had a little surprise for him. It was a thought experiment involving double slits. But these were mounted on rollers or wheels, which made the apparatus movable. Now suppose you choose on a place on the detector above both slits. An electron comes to the lower slit, say. To reach the place you've chosen, the electron which has come through the slit must change its direction, which means that it must change its momentum. Since momentum is conserved, the apparatus acquires this momentum in reverse—it recoils—and this can be detected, since it will move on its wheels. This recoil momentum can be measured. If the electron comes through the upper slit, there will also be a momentum change, but less, since the angle it goes through is smaller. Hence by measuring the momentum recoil of the apparatus, we can tell through which slit it went, so it would appear that in one and the same experimental arrangement we can determine both the particle aspect and the wave aspect of the electron. The wave aspect would be manifest by the intensity patterns of electrons on the detector. But Bohr had an answer. The recoil momentum of the apparatus must be measured to a certain accuracy in order to make this determination. But by Heisenberg's uncertainty principle, the accuracy of the position of the slits is limited. If you work out the details, this inaccuracy is greater than the distance between the first and most prominent maxima in the interference pattern. The slits are effectively as wide as the interference pattern, so there is no interference pattern and Einstein's thought experiment does not work. Sometimes one reads in popular books that this setup shows the electron can be in two places at once. The electron is in a "place" only when you observe it on the detecting screen. If someone tells you that it is also in another place, the proper answer is, "Show me." Recall Kerner's remark I quoted at the beginning "An electron can be here or there at the same moment." Insofar as it is "here or there," it is either one or the

other—not both. The play resumes at the point
where Kerner is saying that waves can turn back
into particles.

> *Blair:* How?
> *Kerner:* Nobody knows. Somehow light is contin-
> uous and also discontinuous. The experimenter
> makes the choice. You get what you interrogate
> for . . ."[10]

Blair's question, "How?" is exactly the kind of
question I asked as a freshman and to which there
is no answer, at least in the quantum theory. Stoppard's quantum theory reminds me of a conversation I had with Marvin Minsky when we were both
undergraduates. He later went on to become one
of the founders of artificial intelligence. He showed
me a drawing he had made of a bird. "This looks
like a bird," he said, "but no bird looks like this."
Stoppard's physics looks like quantum mechanics,
but no quantum mechanics looks like Stoppard's
physics.

Professor Frank died on the twenty-first of July
1966. Later that year there was a memorial for
him at Harvard, and I was one of the speakers. His
widow, Hania—whom he had married in Prague
when she was one of his students—was there. She

had belonged to an intellectual circle which included Kafka. She was a remarkable and very lively woman. I have an indelible memory of calling one night to speak to Professor Frank. Hania answered and in her inimitable accent said, "We are here singing English folksongs and Philipp has gone away."

:: 5 ::

A Measurement

We wish to measure a temperature. If we want, we can pursue this process numerically until we have the temperature of the environment of the mercury container of the thermometer, and then say: this temperature is measured by the thermometer. But we can carry the calculation further, and from the properties of the mercury, which can be explained in kinetic and molecular terms, we can calculate its heating, expansion, and the resultant length of the mercury column, and then say: this length is seen by the observer. Going still further, and taking the light source into consideration, we could find out the reflection of the light quanta on the opaque mercury column, and the path of the remaining light quanta into the eye of the observer: their refraction in the eye lens, and the formation of an image on the retina; and then we would say: this image is registered by the retina of the observer. And were our physiological knowledge more precise than it is today, we could go still further, tracing the chemical reactions which produce the impressions of this image on the retina, in the optic nerve tract and in the brain, and then in the end say: these chemical changes of his brain cells are perceived by the observer. But in any case, no matter how we calculate—to the mercury vessel, to the scale of the thermometer, to the retina, or into the brain, at some time we must say: and this is perceived by the observer. That is, we must always divide the world into two parts, the one being the observed system, and the other the observer. In the former, we can follow up all physical processes (in principle at least) arbitrarily precisely. In the latter this is meaningless. The boundary between the two is arbitrary to a very large extent. In particular we saw in the four different possibilities in the example above, that the observer in this sense needs not to become identified with the body of the actual observer: In one instance in the above example, we included even the thermometer in it, while in another instance, even the eyes and optic nerve were not included. That this boundary can be pushed arbitrarily deeply into the interior of the body of the actual observer is the content of the principle of the psycho-physical parallelism—but this does not change the fact that in each method of description the boundary must be put somewhere, if the method is not to proceed vacuously, i.e.; if a comparison with experiment is to be possible. Indeed experience only makes statements of this type: an observer had made a certain (subjective) observation; and never any like this: a physical quantity has a certain value.

—John von Neumann[1]

The idea that consciousness collapses the quantum wave was originally proposed by John von

Neumann in the 1930s. What took us so long to take this idea seriously?

—Amit Goswami[2]

BY THE TIME I finished taking Philipp Frank's course on the foundations of modern physics in the spring of 1948, one thing had become clear to me: if I really wanted to have any kind of understanding, I needed to learn some mathematics. Professor Frank, after each of his formal lectures, left (as I've said) "a certain interval," which was followed by a question period during which he would have a chance to expand on some things "for those of you who know a little of mathematics." I still remember his solving the elementary differential equation that described the decay of a particle. I had no idea what he was doing or how it led to the exponential answer. I also had no idea what *I* was doing, except that I wanted to learn "a little of mathematics." On how I would use this later in life, I was clueless. I had some friends who were taking the introductory course in the calculus. Indeed, one of them showed me his homework papers. The grader had made some extremely witty remarks. It turned out to have been Tom Lehrer, who became one the foremost musical satirists of the twentieth century.

These friends spoke of "infinitesimals," an essential feature of the calculus, as if they represented the limit of possible human understanding. Grasping them, they said, bordered on the impossible. This worried me, so I went to my freshman advisor and asked if he could look up my mathematical aptitude scores from the tests I had taken to get into Harvard in the first place. He did, and said that he thought it would be all right for me to take the calculus. In the fall of my sophomore year, I duly enrolled. I quite enjoyed the course, infinitesimals not withstanding. What I did not enjoy was freshman physics, which I took as a sophomore. I had by no means committed myself to physics as a major, but was testing the waters. I found the course—which dealt with the basics, such as pulleys and levers, inclined planes, and the rest—exceedingly dull. I realized that Professor's Frank's course was the dessert and that I was now dealing with the entrée. To make matters worse, there were laboratory sessions, which I hated. I have told people over the years that the reason for my problems in the lab was myopia. I could not read a slide rule. But the real reason was that I had no mechanical aptitude. Nothing ever worked right. The whole experience has given me a great deal of respect for experimental physicists—indeed, experimental anything. But after that, I never took another undergraduate course in physics.

I now took math courses, both during the academic year and in the summers. That became my major, and I acquired a tutor—the late George Mackey. Mackey was a wonderful teacher who at the time lived a sort of monk's existence devoted to mathematics. Among other things, he was bringing up to date the mathematical foundations of the quantum theory—a more modern version of von Neumann. I learned from Mackey what sort of mathematics was used in the quantum theory, and headed in that direction. Up to this point, I had never taken a course quantum theory. The Mathematics Department at Harvard usually discouraged Harvard undergraduates from applying to the graduate school. They thought it was a good idea for students to have a new experience. You had to have at least a magna cum laude for them to consider your application. By this time, I had come to like life in Cambridge a great deal—and as I had graduated magna, I applied and was accepted. One of the first things I did was to take the graduate course in the quantum theory.

The course was taught by Julian Schwinger, the

star theoretical physicist at Harvard. He was one of the creators of quantum electrodynamics, for which, along with Richard Feynman and Shin-ichiro Tomonaga, he won the 1965 Nobel Prize in Physics. Schwinger's lectures were a form of theater. He lectured seamlessly, without notes, and you were not supposed to ask questions. The first semester was largely devoted to why the quantum theory was necessary and to things like the uncertainty principle. There was no discussion of the Einstein, Podolsky, Rosen paper; and when Bohm's papers came out, at about this time, I once heard Schwinger say that he thought they were pointless. After this introduction, Schwinger launched into his own formalism for the theory, which left most of the class, including me, rather lost. We used to go down to MIT—where Victor Weisskopf was giving a much more homespun version—to fill in the gaps. After two years I received a master's degree, and now had to think seriously about the future.

By this time I had gotten to know a number of mathematicians, including my fellow graduate students. It had become clearer and clearer to me that I did not have the abilities to do creative mathematics at a level that I would find satisfying. Unless you have been involved in one of these creative disciplines, this is difficult to explain. In my case you could point to my course record, which was very good, and ask if that wasn't proof enough. But I could see that my way of understanding things was different from the way the truly gifted students understood them. I could follow proofs, but for them the proof was almost a secondary matter, since the truth of the theorem was to them self-evident. While I was mired in self-doubt, something happened that settled the matter. Garrett Birkhoff, the chairman of the Mathematics Department, called me in for a chat. He had noticed that I had been taking courses in classical mechanics, methods of mathematical physics, and Schwinger's quantum mechanics. He said it seemed to him that I was more interested in physics than in mathematics, and that I would have to chose. I decided that I would try to switch to physics.

Armed with my transcript, I went to the chairman of the Physics Department and asked to make the switch. He looked over my record and drew the conclusion opposite to that of Birkhoff—namely, I had taken too few physics courses. Apart from freshman physics and my graduate courses, I had taken none. On the other hand, I had excellent grades, so I was accepted with two conditions. The first was that I was to spend the summer working at the Harvard Cyclotron Laboratory, to get some real hands-on experience with experimental physics. The second was that I was to take an oral exam on general physics during the next year. I agreed to both.

The first Harvard cyclotron was completed in 1938. When the atomic-bomb project moved to Los Alamos, the cyclotron was sold to the government for one dollar, with an understanding that the government would help to construct a new one after the war. Indeed, this is what happened, and with the aid of the Navy the new cyclotron went into operation in 1949. By present-day standards, it was almost a toy. It was simple enough that faculty and graduate students could operate it. No ordinary physicist would be allowed to get anywhere near the control room of one of the large modern accelerators. That would be like inviting a flight attendant on a transatlantic airliner to come into the cockpit and fly the plane. My link to the cyclotron was a new assistant professor named Walter Selove. He was doing an experiment in nuclear physics, and my job was to help him. But what I actually did had nothing to do with physics. I piled stacks of lead bricks for shielding and fitted various targets to be bombarded by the protons in the beam. I had no idea what the experiment was trying to prove. In short, while I had a good time that summer, I learned nothing.

The following fall, I studied for my forthcoming oral exam and, strange as it may sound, I had begun work on what became my Ph.D. thesis. I had been able to do this because another young faculty member, Abraham Klein, was engaged in a project which had variants of the kind that made theses. They required not a deep understanding of physics but rather the patience to carry out some fairly involved calculations. I think I was Klein's first graduate student. He and Walter Selove went on to have distinguished careers at the University of Pennsylvania. But there was the oral exam. My performance was distinctly odd. I recall one question in which I was asked to show why a certain hypothetical particle decay could not happen. I gave an arcane argument involving something called "charge conjugation," whereas the argument they were looking for involved the conservation of energy and momentum—something that any real physicist would have thought of at once. I did not exactly fail the exam, but I did not exactly pass it either. They would have been within their rights to ask me to repeat it, but instead they asked me to take a course in modern experimental physics.

The course was given by Robert Pound, a noted experimental physicist and a very nice man. Maybe he had been tipped off as to what I was doing there. He treated me very well, including assigning to me a lab partner named Paul Condon. Condon was the son of the very well-known physicist Edward Condon, so I suppose physics was in his genes. His brother Joe was also a physicist. Paul and I soon worked out an amiable modus operandi. I was allowed to comment but not to touch any of the apparatus. I was also allowed to peek at his lab book, which was a model. Pound would come around to our lab bench from time to time and offer encouragement. One of the experiments we did was to measure the spectrum of radiation in a heated cavity: so-called "black-body radiation." It was the report of new measurements of this spectrum at the dawn of the twentieth century that led Max Planck to introduce the notion of the "quantum"— his term. The only way Planck could explain this distribution was to imagine that the electrons attached to the atoms in the walls of the cavity started vibrating when the walls were heated. These oscillating electrons emitted and absorbed radiation.

To produce his distribution, Plank had to assume that these oscillators had a restricted set of energies. In our language, we would say that the energies were "quantized." Professor Frank once noted that it was like beer being sold in pints and quarts—which are analogous to the quanta. Planck introduced a constant that set the scale of these energies. "Planck's constant" is one of the basic natural units. In our modest experiment, we could actually measure this constant. Einstein transformed Planck's idea in 1905, when he argued that the radiation in the cavity was made up of quanta. Professor Frank said it was like the requirement that whenever you found any volume of beer it was always broken up into pints and quarts. In any event, with the help of Condon, I passed the course and was allowed to complete my thesis. At the time, the philosophical implications of our black-body experiment—and some of the others we did—never occurred to me. We were a living realization of Bohr's basic doctrine about the nature of quantum mechanical measurements.

In our experiment, we never dealt with the actual quantum processes that were taking place within the cavity. What we did was to read various dials and pointers and the like. These dials and pointers were completely classical, which is why we could discuss them in everyday language. Bohr insisted that quantum measurements were always like this. At the end of the day, there are observations that must be described in ordinary language, since this is the only language we know. Some division has to be made between the underlying quantum process and the classical world of the observer. Bohr wrote about this again and again. Here is a typical example:

> This necessity of discriminating in each experimental arrangement between those parts of the physical system considered which are to be treated as measuring instruments and those which constitute the objects under investigation may indeed be said to form a principal distinction between classical and quantum-mechanical description of physical phenomena. It is true that the place within each measuring procedure where this discrimination is made is in both cases largely a matter of convenience. While, however, in classical physics the distinction between object and measuring agencies does not entail any difference in the character of the description of the phenomena concerned, its fundamental importance in quantum theory, as we have seen, has its root in the indispensable use of classical concepts in the interpre-

tation of all proper measurements, even though the classical theories do not suffice in accounting for the new types of regularities with which we are concerned in atomic physics.[3]

In his discussions of the foundations of the quantum theory, John Bell introduced the invaluable concept of FAPP: For All Practical Purposes. For most physicists most of the time, when it comes to quantum mechanics, FAPP is perfectly adequate. We may have to dig deeper if we are teaching a course on it and use one of the modern textbooks that discuss such things as the Einstein, Podolsky, Rosen experiment and Bell's inequality. From a FAPP point of view, a statement like the one by Bohr above seems quite reasonable. But it is the sort thing that drove Bell to distraction and caused him to label Bohr as an "obscurantist." Nowhere in Bohr's writings could he find any clear criteria for making the division Bohr talks about between the quantum and classical realms. Bohr appears to maintain that it is somehow a division between large and small things—dials and pointers versus atoms. But this clearly does not work. Take as an example the thought experiment that Einstein came up with in 1927 to refute Bohr's notion of complementarity, the one with the slits on wheels. It is hard to think of anything more classical. But to resolve the paradox that Einstein presented, one must invoke the uncertainty principle. It is hard to think of anything more quantum mechanical. Where in this experiment has the classical realm stopped? How do we know in principle—and not FAPP—where to make this division? You will not find the answer in Bohr. You will likewise not find any discussion of what exactly a quantum mechanical measurement is. In particular, can you describe one using the formalism you might find in an average textbook?

The first person who seems to have discussed this in any generality was von Neumann, in his 1932 book on the mathematical foundations of the theory. He made a distinction between what he called two types of "intervention" which can occur in a quantum mechanical system. One type of intervention is describable by the ordinary Schrödinger equation. We begin by specifying the Schrödinger wave function at some time and place. The system is acted on by a force—a magnetic field, say. The wave function continually evolves according to the Schrödinger equation, until at some later time and at some other place it acquires a new value. If we wish, we can use these values as the initial conditions and run the scenario backward in time. To use the term of art, the equation is "time reversible." But so far, we have not performed a measurement. To see what is involved, let us consider a specific example: the measurement of spins using magnets.

Suppose we have some source that produces spin-½ particles such as electrons. A beam of them is produced by the source that has a mixture of spin up and spin down, ↑ and ↓, with proportions given by numbers a and b. The spin part of the wave function we can write as $a{\uparrow} + b{\downarrow}$. Until these particles encounter the magnet, the wave function propagates as described by the Schrödinger equation. But we know from experiment that once the beam interacts with the magnet, it is divided into two components, one with the spin up particles and one with the spin down. This process is still reversible. But once a particle in one of the beams is actually detected, the situation is drastically altered: one of the components of the wave function disappears. The question is: Can this actual measurement be described by an ordinary time-reversible Schrödinger equation? The answer is no. Once a particle has been measured in the detector with, say, spin up, its wave function is spin

up. The spin down part is gone. This is often referred to as the "collapse of the wave function." It
seems more like a decapitation. It cannot be described by the usual Schrödinger equation. There
are various ways of proving this, depending on the
level of mathematical sophistication one wants to
bring to bear. The point is that this measurement is
not time reversible. You cannot use the Schrödinger
equation to reconstruct the status quo ante. All the
king's horses and all the king's men will not create
the old wave function again. In fact, you cannot
reconstruct the past—which puts history in an odd
light. All you can do is to offer different pasts with
different probabilities. Von Neumann does not discuss this. He simply states as a postulate that you
need a different kind of intervention to carry out
measurements, and he presents their mathematical
properties. This is the "measurement problem,"
and we have been living with it ever since.

Veritable forests have been cut down to produce
the paper that has been expended on this. I will
simply give a personal overview. I state from the
beginning that I do not know what the solution is. I
will divide my discussion into two parts. The first
part will deal with the class of proposed solutions
in which the wave function does not collapse, and

the second part deals with the class of proposed solutions in which the wave function does collapse and an explanation is offered. Each of these classes has its proponents and critics.

The first example in the first class is Bohmian mechanics. In the first of his two papers, Bohm discusses an example of what is now fashionably called "decoherence."[4] It will be recalled that the Schrödinger equation in Bohmian mechanics is used to derive the guide waves for the classical particles. What Bohm states is that when the entangled system enters a region where there is a perturbation due to the interaction with a measuring apparatus, there is a very rapid exchange of energy between the parts of the entangled function, which causes it to break up into separable wave packets: decoherence. These separated packets guide the particles to the various parts of the detector. The wave function does not collapse, but rather breaks up into pieces. It has been shown that this produces the quantum mechanical answer.

Probably the most popular version of the non-collapsing situation goes under the name "many worlds," "many paths," or "many histories." It dates to a 1957 Ph.D. thesis written at Princeton by Hugh Everett III, under the supervision of the late John Wheeler. The thesis was called "The Theory of the Universal Wave Function." Later in 1957, Everett published a paper on it with the more austere title, "'Relative State' Formulation of Quantum Mechanics," a name suggested by Wheeler.[5] So far as I know, this is the only thing Everett ever published on this subject. Wheeler was, at least at first, very enthusiastic about it. He arranged for Everett to visit Bohr in Copenhagen. Neither Bohr, nor anyone else at the time, showed any interest Everett's interpretation. He then got out of physics and made a great deal of money as a defense analyst. He died of a heart attack in 1982, at the age of fifty-one. By this time, there was considerable interest, spearheaded by the physicist Bryce Seligman DeWitt. In the 1970s he wrote both technical and popular articles about it and coined the name "many-universe" interpretation. It is now more often called the "many-worlds interpretation." What is it?

As in Bohmian mechanics, it says that entangled wave functions decohere when the system encounters some sort of perturbation. This perturbation could be, for example, from the photons left over from the Big Bang. Measurements are just another kind of perturbation. In Bohmian mechanics, the different branches of the wave function act as different guide waves for the classical particles. In the Everett interpretation, a wave function is simply a wave function; and it is a solution, at all places and at all times, of the Schrödinger equation. But we follow only one of the branches. The others do not disappear, but they have nothing to do with us. They belong to another world, with which we have no contact. If we use a magnet to measure a spin and find a spin up, the spin down part of the wave function is not lopped off but represents another possible outcome, which is not our concern.

What has been the concern of the physicists who have rejected this interpretation is, among other things, the proliferation of "worlds." A world tree must acquire so many branches that it begins to resemble a forest. This is an aesthetic objection, but there is also a physics objection: Born's rule. Returning to our spin example, recall that the coefficients a and b in the wave function determine the probability that an experiment will yield spin up or spin down. Experiment confirms this assignment. Any interpretation of the quantum theory must yield this rule. It is all very well to say that we are on one branch or another of the Everett's many worlds. But what is the probability that we find

ourselves on this branch and not that one (Born's
rule)? There has been much work on this matter
by the practitioners, but so far as I can tell, the
derivation of Born's rule in the many-worlds interpretation is still an open question about which
there is much disagreement. I now turn to the other
possibility—that the wave function does collapse.

Whatever method one invents to get the wave
function to collapse must go beyond the ordinary
Schrödinger equation. Some new physics must enter. One possibility is that there is a wave-collapsing
force that acts randomly but just in the right places
and times to do the job without producing conflicts. This has been explored in some detail. It will
be more convincing if some of the novel proposed
observable consequences are ever observed. Another avenue has been the claim that it is human
consciousness that acts to collapse the wave function. Before one dismisses this as so much New Age
flapdoodle, one should at least consider some of the
sources. Von Neumann seemed to hint that this is
what he believed. His long-time friend and fellow
Hungarian Eugene Wigner, who was one of the
most important theoretical physicists of the twentieth century, was absolutely convinced. When he
was awarded the Nobel Prize in 1963, he apparently asked his caller from Stockholm which of his

achievements had caused him to be awarded the prize. I doubt that it was for his 1961 article, which he called "Remarks on the Mind-Body Question."[6]

Wigner makes his position clear right from the outset:

> When the province of physical theory was extended to encompass microscopic phenomena, through the creation of quantum mechanics, the concept of consciousness came to the forefront: it was not possible to formulate the laws of quantum mechanics in a fully consistent way without reference to consciousness. All that quantum mechanics purports to provide are probability connections between subsequent impressions (also called "apperceptions") of the consciousness, and even though the dividing line between the observer whose consciousness is being affected, and the observed physical object can be shifted towards the one or the other, it cannot be eliminated. It may be premature to believe that the present philosophy of quantum mechanics will remain a permanent feature of future physical theories, [but] it will remain remarkable, in whatever way our future concepts may develop, that the very study of the external world led to the conclusion that the content of the consciousness is an ultimate reality.[7]

One may well wonder, if humans had not fortuitously taken over from pterodactyls, whether the collective consciousnesses of pterodactyls would have done the trick. Be that as it may, Wigner goes on to a discussion which has become, in these circles, as celebrated as Schrödinger's cat.

> In general there are many types of interactions into which one can enter with the system, leading to different types of observations and measurements. Also, the probabilities of the various possible impressions gained at the next interaction may depend not only on the last, but on the results of many prior observations. The important point is that the impression which one gains at an interaction may, and in general does, modify the probabilities with which one gains the various possible impressions at later interactions. In other words, the impression which one gains at an interaction, called also the result of an observation, modifies the wave function of the system. The modified wave function is, furthermore, in general unpredictable before the impression gained at the interaction has entered our consciousness: it is the entering of an impression into our consciousness which alters the wave function because it modifies our appraisal of the probabilities for different impressions which we expect to receive in the future. It is at this point that the consciousness enters the theory unavoidably. If one speaks in terms of the wave function, its changes are coupled with the entering of impressions into our consciousness. If one formulates the laws of quantum mechanics

in terms of probabilities of impressions, these are ipso facto the primary concepts with which one deals.[8]

But Wigner goes further. It is this part which has been seized on by the multitudes—Wigner's friend. He writes,

It is natural to inquire about the situation if one does not make the observation oneself, but lets someone else carry it out. What is the wave function if my friend looked at the place where the flash might show at time t? The answer is that the information about the object cannot be described by a wave function. One could attribute a wave function to the joint system: friend plus object, and this joint system would have a wave function also after the interaction, that is, after my friend has looked. I can then enter into interaction with this joint system by asking my friend whether he saw a flash. If his answer gives me the impression that he did, the joint wave function of friend + object will change into one in which they even have separate wave functions . . . and the wave function of the object is ψ_1. If he says no, the wave function of the object is ψ_2; i.e., the object behaves from then on as if I had observed it and had seen no flash. However, even in this case, in which the observation was carried out by someone else, the typical change in the wave function occurred only when some information (the yes or the no of my friend) entered my consciousness. It follows that the quantum description of objects is influenced by impressions entering my consciousness.[9]

Given a passage like this, it is little wonder that many nonphysicists, especially those with a penchant for Eastern mysticism, see the floodgates of oneness opening. They should be warned that Wigner's is a minority view among physicists. Most reject the role of consciousness in quantum mechanics. Things are hard enough without introducing something that is even harder, and less well understood. When I think of physicists who invoke consciousness as an "explanation" of the quantum theory, I am reminded of an old joke. A man goes to a psychiatrist and says, "Doc, my brother needs help. He thinks he's a chicken." "Why don't you show him that he isn't?" the psychiatrist asks. "I would," the man says, "except I need the eggs."

:: 6 ::

Entanglements

We continue. That a portion of the knowledge should float in the form of disjunctive conditional statements between the two systems can certainly not happen if we bring up the two from opposite ends of the world and juxtapose them without interaction. For then indeed the two "know" nothing about each other. A measurement of one cannot possibly furnish any grasp of what is to be expected of the other. Any "entanglement of predictions" that takes place can obviously only go back to the fact that the two bodies at some earlier time formed in a true sense one system, that is, were interacting and have left behind traces on each other. If two separated bodies, each by itself known maximally, enter a situation in which they influence each other, and separate again, then there occurs regularly that which I have just called entanglement of our knowledge of the two bodies.

—Erwin Schrödinger[1]

To summarize, Bell's theorem showed, in 1964, that either the statistical predictions of quantum theory are false or the principle of local causes is false. In 1972, Clauser and Freedman performed an experiment at Berkeley which validated the relevant statistical predictions of quantum theory. Therefore, according to Bell's theorem, the principle of local causes must be false.

The principle of local causes says that what happens in one area does not depend upon variables subject to control of an experimenter in a distant space-like separated area. The simplest way to explain the failure of the principle of local causes is to conclude that what happens in one area does depend upon variables subject to the control of an experimenter in a distant space-like separated area. If this explanation is correct, then we live in a non-local universe ("locality fails") characterized by superluminal (faster than light) connections between apparently "separate parts."

—Gary Zukav[2]

Bell's theorem supports Bohr's position and proves rigorously that Einstein's view of physical reality as consisting of independent, spatially separated elements is incompatible with the laws of quantum theory. In other words, Bell's theorem demonstrates that the universe is fundamentally interconnected, interdependent, and inseparable. As the Buddhist sage Nagarjuna put it, hundreds of years ago, "Things derive their being and nature by mutual dependence and are nothing in themselves."

—Fritjof Capra[3]

IN 1907, EINSTEIN published a paper in the *Annalen der Physik* in which he tied up some loose ends left over from his great 1905 paper on the special theory of relativity—"special" because it did not deal with accelerated motions.[4] I will describe one of the loose ends in more modern language. Suppose that the theory of relativity is valid, but suppose there exist particles that can move faster than light. Such hypothetical particles are now called "tachyons."[5] We must be clear about something. An ordinary particle like an electron— if we assume the theory of relativity is correct—can never be accelerated to the speed of light, let alone go faster. Tachyons always move at speeds greater than light. It turns out that this does not violate the theory of relativity, but it leads to very strange behavior indeed.

Let us suppose that you have a tachyon emitter. At some time t you emit a tachyon, and at a later time it is absorbed by a tachyon absorber. This is a perfectly causal sequence. Einstein considered how an observer in motion with respect to you would see the same sequence. It turns out that there is a speed less than the speed of light at which this observer would see these two events occur simultaneously. At speeds greater than this, but still less than the speed of light, the time order of the events would be reversed. This observer would see the tachyon absorbed before it was emitted; or, equivalently, an anti-tachyon would propagate backward in time. In one of his essays, John Bell used this scenario to create the perfect murder.[6] Suppose I lure the victim to a certain spot—I shall call it the origin of coordinates—at a time that is noon in the rest system. An observer in the moving system— the perpetrator, me—will also call the time noon at the common spatial origin of coordinates. But now what I do is to fire my tachyon gun at some time before noon in my system, so that it arrives at the origin at noon. To me, the sequence is perfectly causal. But if I have adjusted my speed correctly, then to you—the observer in the resting system—it will appear as if at noon the victim dropped dead emitting an anti-tachyon which will be absorbed by my gun at a later time. So far as you are concerned, there was no murder. Ruling out such strange behavior requires an assumption beyond relativity. This is usually called "locality." In local theories, superluminal communication is not allowed. Most physicists would regard nonlocal theories as something of an abomination. I will now begin a discussion of how John Bell became associated with the question of locality or nonlocality, and will start with some biographical information about Bell.

John Stewart Bell was born to Protestant

working-class parents in Belfast on July 28, 1928. He once described his family to me as being "poor but honest." His father, who dropped out of school at age eight, held a variety of jobs, including horse trader. There was no tradition in the family for any sort of higher education. Anything above elementary school in Great Britain at that time was neither free nor compulsory. It was expected that Bell would leave school at age fourteen and begin working to help support the family. But his mother encouraged him to continue his education, and somehow enough money was found so that he could enroll in the Belfast Technical High School—basically a trade school, where he studied such things as bricklaying and carpentry. Yet during high school, Bell became interested in philosophy and began reading what he described as "big books on Greek philosophy." He became disillusioned because he thought that all philosophers did was to attempt to refute each other. He did get a glimmering of physics in his high school and decided that, while physics did not deal with the larger questions of life and death, it did deal with the laws of nature. This is what Bell decided to pursue. But when he graduated, he was both too young and too poor to continue his education.

Bell tried unsuccessfully to find work, until he

hit on a job that influenced the course of his career. He was hired by George Karl Emeleus, a physics instructor at Queen's College, Belfast, to set up laboratory experiments for freshman physics students. Emeleus must have seen something exceptional in Bell, because he also suggested physics books for him to read and enabled Bell to listen to the lectures in freshman physics. After a year (this was 1945), Bell was old enough, and had put together enough money, to matriculate. It was at Queen's College that Bell first began learning about quantum mechanics. What he learned outraged him. He understood the mathematics perfectly well, but he did not understand how it was being interpreted. He had flaming arguments with one of his professors, Robert Sloane. I asked Bell if he thought then that the theory was simply false. "I did not dare to think it was false," Bell replied, "but I knew it was rotten." With his Irish lilt, Bell said "rotten" with considerable relish.

What bothered Bell then, and bothered him for the rest of his life, were questions like: What is the Schrödinger wave function? The standard interpretation tells us that the wave function provides information about a quantum mechanical system, such as the probability that, when measured, "observables" have some particular value. But is this all we have—"information"? Isn't there something else behind the scenes? Bell used to conjecture that—while physicists, when they teach or even use the quantum theory, seem to accept the idea that all we have is information—in their heart of hearts they really believe that the wave function is not the whole story. The information contained in it refers to something else, something not revealed —a hidden reality. The Wizard of Oz may put on a great show, but behind the screen there is an old man from Omaha turning the cranks.

Bell graduated from Queens with first-class honors in physics, and decided that he could no longer freeload off his parents. He had been living at home while he was at Queens. He got a job with the Atomic Energy Research establishment in Malvern, Worcestershire, where there was group designing a linear accelerator. Unlike a cyclotron, where the particles move in a circle, a linear accelerator moves particles in straight lines. There are some advantages to this when it comes to how the particles radiate. In Malvern, Bell met his future wife, Mary, a Scottish physicist who had likewise come there to do research. He also grew a red beard, to cover up a scar acquired in a motorbike accident.

For the rest of their careers, both Bells continued
to work on the design of accelerators. John had no
intention of studying for a higher degree, but in
1953 he was selected for a program of the Atomic
Energy Authority which allowed a year's study toward a Ph.D. John decided to go to Birmingham to
study with Rudolf (later Sir Rudolf) Peierls, who
was one of the great twentieth-century teachers of
physics. By this time, Bell had published papers in
accelerator design, but he had also been thinking
about the foundations of the quantum theory.

While still at Queens, Bell had read Max Born's
book *Natural Philosophy of Cause and Chance.*
Born, who died in 1970, was one of the founders of
the quantum theory, although his work was not
recognized for a Nobel Prize until 1954. He had
escaped from Germany in 1933. He took up a position in Cambridge and then moved to the University of Edinburgh. But in 1948 he delivered the
so-called Waynflete Lectures at Oxford. These became the basis of his book, which was published in
1949. Born presented a masterful survey of all of
twentieth-century physics, including of course the
quantum theory. Born was at odds with Einstein
about the interpretation of the theory. He had proposed its probability interpretation, inspiring Ein-

stein's famous letter to him about God not playing
dice. It was a passage in the lectures that struck
Bell. By 1949 it had become clear that there were
problems in things like quantum electrodynamics.
While this theory made fantastically accurate predictions, quantities that should be finite were infinite in the theory. Born addressed the question of
whether these difficulties could be cured by some
return to an underlying classical physics.

> I expect that our present theory will be profoundly
> modified. For it is full of difficulties which I have
> not mentioned at all—the self energies of particles
> in interaction and many other quantities, like col-
> lision cross-sections, lead to divergent integrals.
> [A divergent integral is infinite.] But I should
> never expect that these difficulties could be solved
> by a return to classical concepts. I expect just the
> opposite, that we shall have to sacrifice some cur-
> rent ideas and use still more abstract methods.
> However, these are only opinions. A more concrete
> contribution to this has been made by J. v. Neu-
> mann in his brilliant book *Mathematische Grund-*
> *lagen der Quantenmechanik.* He puts the theory
> on an axiomatic basis by deriving it from a few
> postulates of a very plausible and general charac-
> ter, about the properties of "expectation values"
> (averages) and their representation by mathemat-
> ical symbols. The result is that the formulation

of quantum mechanics is uniquely determined by these axioms; in particular no concealed parameters can be introduced with the help of which the indeterministic description could be transformed into a deterministic one. Hence if a future theory should be deterministic, it cannot be a modification of the present one but must be something essentially different. How this could be possible without sacrificing a whole treasure of well-established results I leave to the determinists to worry about.[7]

Bell was much taken by this and also by the fact that, in early 1952, David Bohm published two papers which did precisely what von Neumann said was impossible. Before I describe Bell's resolution of this paradox and what it led to, I would like to reiterate what Bohm did. First let me say what he did not do. He did not produce a theory that supplants quantum theory. Bohmian mechanics, as it is often called (although never by Bohm), is a reinterpretation of nonrelativistic quantum mechanics. I will come back to the nonrelativistic issue later. It claims to reproduce all the standard results of the quantum theory in a deterministic way. It is sometimes called a theory of "hidden variables," but this is misleading. There is nothing hidden about the trajectories of the particles in Bohmian mechanics. Each trajectory is explicit and classical. These particles are the primary subjects of the theory. The secondary subjects of the theory are the Schrödinger guide waves that guide them. For the sake of historical fairness, it should be mentioned that, after the invention of the quantum theory, de Broglie had presented a very similar model. He abandoned it after he was subjected to some scathing and ultimately irrelevant criticism by Pauli.

As I mentioned earlier, there are two basic equations in the theory. There is the Schrödinger equation, which determines the guide wave. Once you use this equation to solve for the wave function, you can use this solution in a very classical-looking equation for the particles—an equation that determines their trajectories. Since these trajectories are perfectly classical, one may well wonder where, in all this, one can find such quantum mechanical necessities as the Heisenberg uncertainty principle. While it is true that any solution of the Bohmian Schrödinger equation determines a unique trajectory of the particles with no uncertainties, the initial conditions that determine the wave function are uncertain, and this is reflected in the trajectories. This is reminiscent of some of the discussions of determinism that arose after the uncertainty principle was discovered. It used to be said that the Newtonian universe evolved like clockwork. Once the initial conditions were given—the positions and momenta of all the particles—the future was completely determined. But if you could never precisely determine these initial conditions, as Heisenberg's uncertainty principle said, then you were liberated from this deterministic straitjacket. I leave it to the philosophers to elaborate this discussion. In the case of Bohmian mechanics, it is the uncertainty in the initial conditions that generates the uncertainties in the trajectories. One might object that this is something that is imposed on the theory and not, as in ordinary quantum mechanics, derived from it. The real question is: Is the game worth the candle? Is trying to resolve some unease about the interpretation of the quantum theory worth all the trouble of Bohmian mechanics?

To most working physicists, the answer is very probably no. Just as one can successfully ride a bicycle without delving into the gyroscopic principles that govern turning, one can use the rules of quantum theory to make calculations and predictions without worrying a great deal about what it all means. This has changed to some extent since Bell's work, and many distinguished physicists now feel

it is legitimate to look into these matters. Bell was
such a good physicist, that people felt: If he can do
this, so can I. Bohmian mechanics is only one avenue. There are others. Take the matter of the double slit, which caused me so much bewilderment
when I was a college freshman. Electrons, say, impinge on a barrier perforated by two closely spaced
slits. There is a detector behind the barrier for the
electrons. If the intensity of electrons is reduced
so that they enter the apparatus singly, nonetheless, over the course of time, they build up at the
detector, reproducing the wave pattern which a
wave would have produced if it had gone through
the slits. If you close one of the slits, the pattern
changes. The question I asked myself all those
years ago was: How does the electron "know" that
one of the slits has been closed when it passes
through the other? Ordinary quantum mechanics
does not really answer this question. In fact, it dismisses it. There is a rule that tells you how to com-
pute the probability that the electron will land in
one spot or the other on the detector, and that is
pretty much that. In Bohmian mechanics, there is
an answer. If both slits are open, the guide wave
goes through both of them. The interference of the
waves is reflected in the trajectories of the elec-

trons, which go through either one slit or the other. There is something satisfying about this—but there is another problem that is inherent in the whole scheme and that was first clearly presented by Bell. I will get to it after I give a bit more of his biography.

When he went to Birmingham for his sabbatical in 1953, Bell had already published some papers on accelerator design. Because of this, Peierls told Bell that he would not be treated as a "beginner" but was expected to give a seminar on his work. Bell gave Peierls two choices of subject: accelerator design or the foundations of the quantum theory, which would deal with Bohmian mechanics. Peierls made it very clear that he did not want to hear about Bohmian mechanics, for which he had no use. Peierls always maintained that there were no problems to be solved in the interpretation of the theory. Much later he did study and admire Bell's work, though he still took the position that the "problems" had to do with muddled thought. Bell did his Ph.D. thesis under the supervision of a young assistant to Peierls named Paul Matthews, who went on to have a very distinguished career. After his sabbatical year, Bell went to Harwell, another station of the Atomic Energy Authority. It had a small group which was devoted to basic research in nuclear physics. He remained there for a few years, and then decided that the group did not really fit into anything else the laboratory was doing, so he should try to relocate. As it happened, he had already consulted on the design for an accelerator that was being built at CERN in Geneva. He and Mary were offered jobs there in 1960, and they spent the rest of their careers at CERN.

Bell had a very strong work ethic. He felt that he was being paid by CERN to do accelerator design and elementary-particle theory, and not to study the foundations of the quantum theory. So that is what he did. But he could not help thinking about the quantum theory. When he was still at Harwell he had a colleague named Franz Mandl, who was a native speaker of German. This was important, because there was no English translation of von Neumann's book then available. Once Mandl translated the relevant part of the book, Bell saw at once what the problem was with von Neumann. Von Neumann had assumed that the average of the sum of quantities is the same as the sum of the averages of them. This is almost trivial for averages in conventional quantum theory. Von Neumann had assumed that it would necessarily apply to any deterministic alternative. Yet Bell realized that it did not apply to Bohmian mechanics, so that von Neumann's argument was irrelevant. He let the matter rest until 1963, when he and Mary got a sabbatical leave from CERN, part of which they spent at Stanford. He felt that he was now free to work on anything.

Bell first published a review article which he called "On the Problem of Hidden Variables in Quantum Mechanics."[8] This was a summary of all the work he had done on the subject since 1952. Most of the article deals with the question of why various proposed "proofs" such as von Neumann's do not rule out Bohm-like interpretations of the quantum theory. But at the end, Bell raises an issue that was only briefly alluded to in Bohm's papers. Suppose you have two Bohmian particles in interaction. This gives rise to two equations of motion, one for each particle. But in general the two particles get intertwined in these equations. The trajectory of one depends on instantaneous knowledge of the behavior of the other, even if the two particles are spatially separated. In short, the theory is nonlocal. It reproduces the results of the quantum theory but at a very heavy cost. Bell then adds a sentence: "However, it must be stressed that, to the present writer's knowledge, there is no proof that any hidden variable account of quantum mechanics must have this extraordinary character."[9] To this is appended a footnote that reads, "Since the completion of this paper such a proof has been found (J. S. Bell, *Physics*, 1 [1965], 195)." This is a reference to the paper in which Bell proved his theorem. Because of this work and what it led to, Bell had been nominated for the Nobel Prize by the time of his death on October 1, 1990.

To understand what Bell did, we must back up to 1910. This was the year that the New Zealand–born physicist Ernest Rutherford, and his young associates at Manchester, discovered the atomic nucleus. Rutherford then presented the model of the atom which has been with us ever since. Here is its modern version. There is a tiny massive nucleus made up of neutrons and protons which is surrounded, and at a great distance from the nucleus, by a sufficient number of negatively charged electrons so as to balance the charge of the protons. When in 1911 Niels Bohr came as a postdoctoral fellow from Denmark to work with Rutherford, he had available to him this model of the atom, as well as the puzzle it posed. What kept the whole thing stable? What prevented the electrons from just collapsing into the nucleus, considering that they were electrically attracted to the protons? After his return to Denmark, Bohr proposed an answer, which he published in 1913.

In his model, the electrons are restricted to certain orbits. In our language, the orbits are "quantized." The orbit with the least energy—the "ground state"—is, by hypothesis, stable. But electrons in the ground state can be excited to states of higher energy. When they return to the ground state, radiation is emitted as the electrons transition from the states of higher energy to those of lower energy. The energies of this radiation are determined by the energies of the electrons in these quantized orbits. Indeed, the difference in energy between one orbit and another is the energy that is radiated. This means that, instead of a chaos of emitted radiation, only certain frequencies are observed. This explains why collections of excited atoms emit the beautiful spectra of light that characterize these atoms like fingerprints.

For the next decade or so after Bohr published his work, there was practically a cottage industry of physicists applying it to more and more complex spectra. All of this—a mixture of classical and new physics—came to be called the "old quantum theory." By the 1920s there were some ineluctable problems which the old quantum theory could not solve. The atomic electron states were characterized by so-called "quantum numbers." In addition to the energies, the electrons in these states carried an orbital angular momentum. This angular momentum was "quantized," meaning that not only did the total angular momentum have a limited number of values it could take on (in contrast to the case in classical physics), but, for a given value, there were only a limited number of directions to which the angular momentum could point. These added quantum numbers gave rise to additional spectral lines, since there were more ways that the electron could make transitions. But there was a problematic doubling of some of the lines that nothing seemed to explain.

Sometimes in physics, when it comes to a conundrum like this, it takes young physicists with nothing to lose to resolve it. In this case the two young physicists were Dutch graduate students: George Uhlenbeck and Samuel Goudsmit. Goudsmit knew a lot about the spectra, and Uhlenbeck had a deeper understanding of theoretical physics. In 1925, they proposed that the electron had an additional angular momentum that had nothing to do with its motion. This became known as "spin." In their original picture, they saw the electron as a tiny ball of charge spinning like the earth on an axis. But this homey picture is untenable. The electron would have to spin so fast that the exterior would be moving faster than the speed of light. Moreover, to fit the data, the electron can spin pointing in only two directions—"up" and "down." Spin is a quantum mechanical property intrinsic to any particle. The introduction of spin resolved the doubling of some of the spectral lines. In fact, although Goudsmit and Uhlenbeck did not realize this, there were data from an entirely different source: an experiment that had been done in 1922 provided conclusive evidence for the existence of spin.

When the German physicist Otto Stern returned from his service in the First World War, he became an assistant to Born in Frankfurt. Born had become interested in what are known as molecular-beam experiments. Molecules are heated in a furnace and then allowed to escape into a directed beam. These molecules are electrically neutral, so they do not get perturbed by stray electric fields that could overwhelm any subtle effects. Born put Stern onto working with these beams. Stern acquired a collaborator named Walther Gerlach, a highly skilled experimental physicist. Stern wanted to test a prediction of the old quantum theory. This had to do with the different directions in which the angular momentum of the molecule could point. Stern chose the silver molecule because it had one electron outside a closed core of electrons.

The total angular momentum of the atom is composed of the angular momentum of the electron and the angular momentum of the core, neither of which was known. But this angular momentum is associated with an atomic magnet whose orientations are those of the total angular momentum. One has, in effect, a tiny magnet that can point in a limited number of directions reflecting the quantization of the atomic angular momentum which is unknown. If this tiny magnet interacts with a strong magnetic field which, in turn, points in some fixed direction, then, depending on how the angular momentum of the atom is oriented, the molecule in the beam would follow one trajectory or another in the external magnetic field. By collecting these atoms on a detector, one could see from these depositions the effect of these different trajectories—how many lines were produced on the detector.

The main thing that Stern wanted to test was the quantization of the angular momentum, which Bohr had predicted. If the angular momentum was classical it could point in any direction, so on the detector there would be a smear. In fact, what turned up were two distinct lines. This showed that the angular momentum was quantized, but the fact that there were only two lines was mysterious. It took five more years before this observation was connected to electron spin. The reason was that, until 1927, it was not known that the core of electrons in the silver atom carried no angular momentum—so that the entire angular momentum of the atom was that of the exterior electron, which had no orbital angular momentum. Therefore the whole effect was due to electron spin. There were two possible trajectories in the Stern-Gerlach magnet because the spin of the electron could point only up or down.

I bring all this up because Stern-Gerlach-like magnets play an essential role in Bell's analysis. It all began with Bohm's text *Quantum Theory*, which he published in 1951. One of the things that Bohm presented was his version of the Einstein, Podolsky, Rosen paper. They had considered wave functions that represented entangled momenta, which made their arguments rather complicated. Bohm focused on spin, which could point only up or down.

In the units in which spins are measured, the electron has spin ½, as do the neutron and proton. If we put two electrons together, then their net spin can be 1 or 0. For his purposes, Bohm considered the spin-zero state, which is known as the "singlet." To get zero net spin, one of the electrons must have spin up and the other spin down, but in the absence of measurement we do not know which is which. In an obvious notation, the wave function has the form $(1) \uparrow (2) \downarrow - (1) \downarrow (2) \uparrow$. The minus sign is essential to make the net spin zero. This wave function represents the "entanglement" of the two electrons, and indeed, in many ways, it is the paradigmatic example of entanglement.

Suppose there was a particle of spin zero that could decay into two electrons. These electrons would be created in the singlet state. If the particle was at rest at the time of the decay, then the electrons would fly off in opposite directions at the same speed, so as to conserve momentum. Quantum mechanics tells us that, unless one of the electrons interacted with something, the spin entanglement would endure no matter how far apart the electrons were.

Now suppose that I have a Stern-Gerlach magnet somewhere. (The actual Stern-Gerlach experiment was done—for technical reasons, as I have mentioned—with silver atoms and not with free electrons.) When one of the electrons enters the magnet, it can potentially have spin up or spin down. Which one it has is revealed by detecting its trajectory—where it arrives at the detector. Before this detection, all one can say is that a given electron has the potentiality of having spin up or down. If we have a collection of these decaying particles, they will keep supplying electrons to the magnet. Half the time these electrons will be spin up, and half the time spin down. If I am doing the detection at one of the magnets, the distribution of spin up and spin down will appear entirely random. Suppose I note these different measurements in my lab book with a "plus" sign for spin up and a "minus" sign for spin down. Then after a while the entry will look something like $++-+-+-+++$ . . . and so on.

Now, you likewise have a Stern-Gerlach magnet, which is farther away from the electron source in the middle than mine, but oriented in the same way that mine is. Your electrons arrive a little later than mine, and you also have a lab book in which you note the pluses and minuses. To you, they also seem entirely random. Unless we compare notes, this is what we will conclude: a random distribution of spins at each magnet. But later on, we do compare notes. If we know our quantum mechanics, we will not be surprised at the result. The spins measured at our two magnets are perfectly anti-correlated. Whenever I note a plus, you will note a minus, and vice versa.

If I put myself in the state of mind I was in when I was a college freshman first learning about the two-slit experiment with the single electrons, the question I would ask is: How does the distant electron "know" that its mate has just had its spin measured, which forces the anti-correlation? In standard quantum theory with no superluminal signals allowed, there is no answer. It is not even a question. The theory tells you how to calculate the correlation, and that is that. To console yourself, you might think of the following situation. I have a large coin which someone has sliced into two pieces, one with heads and one with tails. This person has put the two halves in sealed bags without telling me or you which is which. I give one of the bags to you and you go off to Kathmandu. There is a fifty-fifty chance that when I open my bag the coin will have the head. I do so, and lo and behold I do have a head. At that instant, the probability that you have a head drops to zero. There is nothing mysterious about this. Probability reflects the state of knowledge, and when this changes so do the probabilities.

The difference between this and the quantum mechanical situation is that if I find a head, I can be sure that it was always a head. I do not have to do a measurement. But in quantum theory, I cannot make that assumption. If I do, then I can generate contradictions with the uncertainty principle. This is the sort of thing that troubled Einstein. He called this sort of quantum mechanical correlation a "spooky" action at a distance. Bohmian mechanics does provide an explanation. When the two particles are entangled, whatever one does influences the behavior of the other instantaneously. In the paper in which Bohm discusses this, he claims that there is no contradiction with relativity because no "information" is being transmitted this way. Perhaps—but what about causal sequences as viewed from different reference frames? It seems to me that once you open the Pandora's box of this kind of instantaneous, nonlocal influence sharing, no one can be sure what is going to pop out.

Bell understood all this. The question he asked himself was: Can you have the quantum mechanical correlations generated by some classical hidden-variable theory, unlike Bohm's, with only local interactions? In thinking about this, Bell proposed a variant of the two parallel Stern-Gerlach magnets that no one had thought of. Suppose you rotate one or both of the magnets. What happens to the correlation? Quantum mechanics has a definite answer. If the two magnets are parallel, then there is the perfect anti-correlation we just discussed. If they are perpendicular, there is no correlation at all. In between, when the magnets are rotated at an angle a, the correlation is given by $-\cos(a)$ which for $a = 0$ and $a = 90°$ gives you the results I just mentioned. The question became: Could any local hidden-variable theory reproduce this correlation? What Bell meant by "local" was that the value of the hidden variables at one magnet did not in any way depend on what was happening at the other magnet. Bell found it easy to exhibit a toy model with these properties for the cases of zero and ninety degrees. But for the angles in between, no such model worked.

Bell's general argument is rather abstract and

sophisticated. I would like to present an analysis which is neither. I don't claim any rigor for what follows. I can tell you that it was suggested by some remarks of Bell's, although without the doodads. Imagine that I have available to me some incredibly clever automata, which I will call "Einstein robots"—the doodads. I can program the robots to do anything except send signals to each other at superluminal speeds. The question is: Can I program them to reproduce the quantum correlation without quantum mechanics? I present one of the robots with a batch of diatomic molecules, each of whose atoms has spin ½. I instruct the robot to split the molecule and fling the components in opposite directions. I tell it to do this repeatedly, making sure that the spin ups and spin downs are opposite in each case, but without prejudice as to which of the components has which spin orientation. You may ask: Where is the entangled wave function? There isn't one—which is just the point. I am doing quantum mechanics without doing quantum mechanics.

To each of the components I attach an Einstein robot. If the Stern-Gerlach magnets are parallel, the robots are instructed to guide the trajectories of the particles suitably, according to whether the spin is up or down. This will reproduce the anti-correlation. Now I rotate one of the magnets through a small angle a. I instruct the robots, when they encounter the rotated angle, to adjust the trajectories so as to reproduce the small-angle correlation $-1 + a^2/2$.[10] So far, so good. I return this magnet to its original position and rotate the other magnet through the angle $-a$. With the same instructions to the robots, I get the same correlation. But now I pull a dirty trick. I rotate one magnet through a and the other through $-a$. The poor robots, which are not allowed to communicate, do not know how to do anything but produce the sum of the correlations for a and $-a$, which is $-1 + a^2$. But the quantum mechanical correlation involves the angle $2a$, which is the angular separation of the two magnets. This correlation is $-1 + \frac{1}{2}(2a)^2$. If we take, say, $a = \frac{1}{2}$, then the classical answer is $-\frac{3}{4}$ while the quantum answer is $-\frac{1}{2}$.

This inequality is a very special case of what is known as Bell's inequalities. What Bell had shown is that no local hidden-variable theory can reproduce all the results of quantum mechanics. Bohmian mechanics can reproduce them because it is not local. It solves the problem of an explanation in the way that Einstein would probably have liked least.

It replaces the "spooky actions at a distance" of quantum mechanics by a theory that has superluminal influences.

According to some theorists, Bell proved that Bohr was right and Einstein was wrong. Bell even said as much, with some regret, since he felt that Einstein was asking the natural scientific question and Bohr was replying by saying that the question should not be asked. Bell found much of Bohr's writing on the quantum theory to be very obscure, which irritated him no end. But it is not clear to me that this result would have displeased Einstein. What we do know is that he did not think much of Bohmian mechanics. He wrote a letter to Born in May 1952 in which he said, "Have you noticed that Bohm believes (as de Broglie did, by the way, twenty-five years ago) that he is able to determine quantum theory in deterministic terms? That way seems too cheap to me. But you, of course, can judge this better than I."[11] I have never understood what Einstein meant by "too cheap," or exactly what he was trying to do. But I am not sure that Bell's results would have made much difference to him.

Having answered the question he set out to answer, Bell had to decide where to publish it. The natural place would have been one of the trade journals, such as the *Physical Review*. But at the time, these all had substantial page charges, which were paid by the entity or individual who had submitted the paper, usually out of government grant money. Bell had no intention of paying these page charges himself, nor did he think it was right to have Stanford pay them, since he was a guest. But it turned out that a new journal named *Physics* was actually paying for articles. This is where Bell sent his paper, which was entitled "On the Einstein Podolsky Rosen Paradox." It was published in 1965 in the first issue.[12] The journal, which rather soon went out of business, did not supply free reprints, so Bell used most of his honorarium from the journal to buy reprints. He might have saved his money, since for the next five years he had almost no requests. There was essentially no interest in what he had done, and he went back to doing elementary-particle theory and designing accelerators. Then he received a letter from a young physicist at Berkeley named John Clauser, and the letter changed everything.

Clauser had taken his degree at Columbia University in experimental physics. Columbia had a requirement that a graduate student get at least

a B in an advanced quantum mechanics course.
Clauser kept getting C's because he could not understand what the quantum manipulations meant.
This led him to read Bohm's papers, and ultimately
to study Bell's work. There was no one at Columbia who had any interest in these things whom he
could talk to. He finally passed the course and got
his degree, and accepted a job at Berkeley working
in the laboratory of the Nobelist Charles Townes.
By this time, Clauser had seen how to test Bell's
inequalities experimentally, and Townes was suffi-
ciently interested to allow him to use the facilities
of the laboratory and to assign him the part-time
assistance of another laboratory physicist, Stuart
Freedman. Clauser's important innovation was
that he did not use electrons in the experiment,
since fact that they were electrically charged would
have overwhelmed any subtle effects. Instead, he
used light.

Light quanta—photons—have spin. In this case,
spin one. Hence, if you produce a pair of entangled
photons, you could in principle measure the correlations of their spins. Clauser derived a new version of the Bell inequality that applied to this case,
and that also took into account some of the limitations of an actual experiment. Unknown to Clauser,

on the East Coast a group of physicists had done the same thing. They were led by the philosopher-physicist Abner Shimony of Boston University, working with his graduate student Michael Horne and a Harvard physicist named Richard Holt. The East Coast and West Coast physicists decided to collaborate, and they published a joint paper in 1969.[13] It has become a standard reference in the field.

Meanwhile, using these ideas, Clauser and Freedman were preparing their experiment. The first problem was: Where do you get correlated photon pairs? One idea was to use the annihilation of the electron with its anti-particle, the positron. This does produce a correlated photon pair—but positrons are hard to come by in quantity, so they settled on an atomic process using cadmium. By 1972, they were able to announce their results, which agreed with quantum mechanics. Then the experiments were taken up by a French physicist named Alain Aspect. His very high-precision results definitely agree with the quantum theory. To most physicists, this settles in the negative the question of whether a local, hidden-variable theory can reproduce the predictions of the quantum theory.

Bell was, of course, following these developments with great interest—but so were a number of other people, many of whom were not physicists. Bell could easily have become a guru, the way Bohm did. But this was not in his nature, so he watched all of this with what I would describe as stoic amusement. He gave many popular lectures. He found a device for leading people into the subject: identical twins who had been separated at birth. Bell discovered that there were a number of pairs that had been reunited later in life to find that they had a remarkable number of things in common, including smoking the same brand of cigarette. We have an explanation for this: they share the same genes. But what is the explanation for the correlations of the widely separated entangled photons or electrons? If you accept the usual interpretation of the quantum theory, there isn't one.

By now, Bell's theorem seems to have gone beyond physics. From time to time, I check some of the tens of thousands of websites of every description on which the theorem is discussed. It is a very mixed bag. Here are two examples. There is a charming site called Kids.Net.Au.[14] One sentence reads, "In 1964 [Bell] demonstrated that quantum mechanics requires superaluminal [?] signaling." The question mark is in the original. Then there is a site that is called "Shamanic Healing: Why It Works."[15] Here are a couple of quotes: "Strained by the conflicts between Einstein and Bohr over the ultimate meaning of quantum mechanics, subjected to further stress in Bell's theorem, and finally ripped through in recent tests, the whole cloth of the materialistic picture of reality must now be rejected." And: "This means that shamanism finally has an explanation based in modern physics. Shamans can effect change in local reality through spirit helpers working at the quantum level. This is achieved through their ritual action, in which the shaman's consciousness, in an altered state of being, is intently focused on a singular objective. For example, 'Take this cancer out of this sick person.'"

I once asked Bell what he thought the problem with the quantum theory was. He laughed and said that if he knew, he might make some progress toward solving it.

:: 7 ::

Anyway, What the #$*! Do We Know?

> It's only in conscious
> Experience that it seems that
> We move forward in time.
> In quantum theory.
> You can also
> Go backwards in time.
>
> —Dr. Stuart Hameroff[1]

Quantum mechanics allows for the intangible phenomenon of freedom to be woven into human nature.

> —Dr. Jeffrey Satinover[2]

But it had also been Schrödinger who had convincingly argued that because the wave function is stubbornly smeared out over configuration space, the abstract space of all possible configurations of particles, until the precise moment of its collapse, it therefore resists all attempts to connect up with a world recognizably like our own. Faced with this intractability, formally known as the "measurement problem," many of the luminaries of physics, from Bohr and Heisenberg on down, took a radical step of denying the existence of an independently existing world altogether and, surprisingly, got away with it. In other, i.e. nonscientific contexts, the difference between those who are committed to an independently existing reality, and those who are not, is roughly correlated with the distinction between the sane and the psychotic.

—Rebecca Goldstein[3]

When he first met Desplechin in 1982, Djerzinski was finishing his doctoral thesis at the University of Orsay. As part of his studies he took part in Alain Aspect's groundbreaking experiments, which showed that the behavior of photons emitted in succession from a single calcium atom was inseparable from the others. Michel was the youngest researcher on the team.

—Michel Houellebecq, *The Elementary Particles*[4]

Where there's smoke, there's smoke.

—John Wheeler

I WENT THROUGH A chess-playing phase. I was never so good that it became an obsession, but I did study and take a few lessons. I played chess hustlers—and still do—in Washington Square Park in New York's Greenwich Village. The first time I met Stanley Kubrick I discovered that he had been, somewhat before my time, one of those chess hustlers. This led to a match of twenty-five games—

which he won—while he was filming *2001*. My account of this in the *New Yorker* led to a commission from *Playboy* magazine to cover the 1972 match between Bobby Fischer and Boris Spassky in Iceland. I tell you this because of a missed opportunity. One time when I was visiting my parents in Rochester, N.Y., I learned that the International Grandmaster Samuel Reshevsky was giving an exhibition. Reshevsky had dominated American chess for several decades and was one of my heroes. He was going to play a simultaneous match with some forty players, each one paying a few dollars for the privilege. I signed up.

In an exhibition like this, the boards are set up on the periphery and the master makes the rounds from the inside. Reshevsky stopped at each board for what seemed like mere seconds before he made his move and went on to the next player. This gave one some time to contemplate one's next move and to study the other participants. I noticed, a few boards from mine, a man with a very distinctive angular face. I did not know who he was, but he was certainly not from Rochester. Later, when I asked, I was told that it was Marcel Duchamp. By 1923, he had essentially given up art for chess, and in 1925 he came close to winning the French championship. He designed and built most of his own set carved from wood, although a local craftsman did the knights. His wife at the time was not amused and glued the pieces to the board. The marriage did not survive much longer. The missed opportunity? I did not realize that Duchamp had read both Einstein and Henri Poincaré on relativity. How much this influenced a painting like *Nude Descending a Staircase* is much discussed. I could have asked him.

While there does not seem to be a consensus about the influence of relativity on Duchamp, there is no doubt about its influence on Lawrence Durrell. In an interview with Dieter Zimmer he said, "The first three parts of my [*Alexandria*] *Quartet* develop the three dimensions of space, each from a different point of view or dimension so that they come in conflict with each other. The fourth then adds time to the dimensions of space. You see, there is a parallel to the theory of relativity."[5]

It is said that quantum mechanics played a role in Durrell's *Avignon Quintet*. I don't see it, but I may have missed something. It is also said that quantum mechanics plays a role in Thomas Pynchon's novel *Gravity's Rainbow*. Pynchon did briefly study engineering physics at Cornell. There is real science in *Gravity's Rainbow.* Here is an example. One of Pynchon's characters is called Roger Mexico. He is a statistician. He decides to apply statistical theory to the explosions of the V2 rockets fired by the Germans toward London. He divides London into squares and plots the number of explosions in each square. He discovers that they follow what is known as a Poisson distribution: the landings are by chance. This means that whatever guidance system the Germans had on their rockets, while it was good enough to target London, it was not good enough to target, say, Buckingham Palace—a useful piece of information. Remarkably, there was a real-life British actuary named R. D. Clarke who did just this. He published a one-page paper on it in 1946. Pynchon must have seen it, because Clarke and Mexico use the same mathematical notation. This is not quantum mechanics, but it is fascinating nonetheless. There are two novelists who do use quantum mechanics in their novels: Rebecca Goldstein and Michel Houellebecq. First Goldstein.

She was born in 1950 into an orthodox Jewish family and grew up in White Plains, New York. She did her undergraduate work at Barnard College and earned a Ph.D. in philosophy at Princeton. She then taught at various universities and won a Mac-Arthur grant in recognition of her writing. Her teaching has given her a great deal of opportunity to study academic life, which, in my experience, is often governed by the partial obverse of Lord Acton's maxim: the absolute lack of power corrupts absolutely. Tenured faculty, who have come to despise each other, have only the power to make each other's lives as miserable as possible. In Goldstein's novel *Properties of Light*, where quantum theory plays an essential role, these characteristics are in full flower.

Much of the action in the novel centers around the Physics Department of a prestigious eastern university. Princeton? The department chairman is an odious man named Dietrich Spencer, and one of the most senior physicists is a man named Samuel Mallach. In an afterword, Goldstein tells us that the character of Mallach was inspired by David Bohm, whom she knew. She has changed many of the details. Mallach has a beautiful daughter named Dana; Bohm was childless. Mallach's wife was an alcoholic; Bohm's wife, Sarah, wasn't. Mallach's father sold used furniture in Scranton, Pennsylvania; Bohm's father sold used furniture in Wilkes-Barre. A similarity: both fathers were devoutly Jewish, yet neither son was. Of much greater significance is the fact that both men wrote papers on a hidden-variable deterministic interpretation of the quantum theory. The interpretation has caused Mallach to be exiled by the members of his department, who dismiss his work as useless, or worse. He has been assigned to teach in perpetuity a course for nonphysicists called Physics for Poets. In my experience, there are not many poets who take this course if they can avoid it. Mallach has come to love the poetry, but he is an abominable teacher whom his students despise. He, in turn, despises—hates—Spencer and comes to the department only when he has to teach his course. He has completely abandoned physics and can hardly remember the content of his own papers.

Into this witches' cauldron is injected Justin Childs. He is an extremely brilliant young theoretical physicist who has learned his quantum mechanics at Paradise Tech in California. He has no problem at all with the formal aspects of the theory. It is its meaning that he has trouble with. Then he comes upon the paper by Mallach. Goldstein writes:

> Justin had read Mallach's paper with something like amazement, for in it he saw clearly that

the impossible had been done: an objective model for quantum mechanics.

Mallach had formulated a hidden-variable version of quantum physics that had accomplished wonders for the material world, saved it from the mathemysticians, the kabbalists of Copenhagen, waving their hands and intoning their obfuscations, turning matter into a miserable ghost of itself, suspended in the not-quite-here-not-quite-there quantum paradox, in which mess they reveled. Mallach's work, Mallachian mechanics as Justin would eventually dub it (although not without demurs from Mallach himself), was the very counter charm to break the vicious spell.[6]

I was reminded of John Bell's discovery of Bohm at about the same point in his career. He too had been dissatisfied with the "kabbalists of Copenhagen" and had learned that the great von Neumann had "proven" that a hidden-variables theory could not reproduce quantum mechanics. Then in 1952 came the papers by Bohm. Bell writes, "In 1952 I saw the impossible done. It was in the papers by David Bohm. Bohm showed explicitly how parameters could indeed be introduced into non-relativistic wave mechanics, with the help of which the indeterministic description could be transformed into a deterministic one. More importantly, in my opinion, the subjectivity of the orthodox version, the necessary reference to the 'observer,' could be eliminated."[7]

As I have mentioned, by 1952 Bohm had left the United States for Brazil. Investigations were weighing down on him because of his Communist background, so Bell could not have consulted him then if he had wanted to. On the other hand, Mallach is in a nearby office so Justin Childs has no problem finding him. At first Mallach thinks that Justin is a student. No member of his department comes to consult him about anything. When he realizes that Childs is a faculty member, he says he would prefer that they go to his house. When they get there, Mallach takes Childs to his study and has to find an additional chair. There ensues an odd monologue in which Mallach says things like, "Perhaps . . . what we learn from the wave function is nothing of the system but only something of the systematizer, just as dream descriptions are only revelatory of the dreamer."[8] This goes on for a while, until Mallach says to Childs, "I do not know what it is you want from me. I cannot fathom what it is you are after. Why don't you go speak to my daughter? I have a daughter, you know."[9] Enter the beautiful Dana.

Dana is attracted to Childs and seduces him on the spot. It does not seem to surprise, or even much interest, Mallach that Childs appears at breakfast the next morning. Much of the remainder of the book is devoted to this entangled and ultimately tragic love affair. One of the things that holds it together is that the three of them embark on a project to fill in the lacunae in Mallach's paper. Each of them brings a different skill set to the task. Mallach brings his original insights; Childs brings his prodigious mathematical skill; and Dana, while she has less formal training than they do, has a phenomenal instinct for the physics.

A reader of the novel does not really need to know quantum theory. In fact, one might be better off if one does not understand the issues and simply takes the theory as metaphor. I am, fortunately or unfortunately, not in this position. While I was reading what Goldstein writes about their project, I kept thinking of something that Wolfgang Pauli once said in a similar situation. I offer it to you *toute proportion gardée*. "I know much," he said. "I know too much. I am a quantum elder." Here is one of Goldstein's explanations:

> Mallachian mechanics seems, at first blush, more irreconcilable with relativity theory than other formulations of quantum mechanics, as the very few physicists who had bothered to take notice of his work emphasized. But this is only be-

cause it brings into the light what the other for-
mulations obscure, which is the utterly startling,
but nonetheless utterly undeniable fact that nature is, in the word of the physicist, "nonlocal":
events can have instantaneous influences on other,
far-distant events. A crude way of seeing that nonlocality—and therefore quantum mechanics—is
at odds with Einstein's relativity theory is to remember that it is fundamental to relativity that
nothing can travel faster than light, whereas these
instantaneous propagations of influence seem to
indicate superluminal, even infinite velocities. The
cognitive dissonance can be expressed in more
subtle and technical terms, involving the relativistic conditions of Lorentz invariance, which, when
combined with quantum nonlocality, allows for
such unacceptable anomalies as "backwards cau-
sation," which would be the future affecting the
past.[10]

The first thing that struck me about this remarkable paragraph was the assertion that relativity and quantum theory are "at odds" with each
other. Even Einstein never made a claim like this.
Trying to interpret the phrase "at odds," I can only
assume that the implication is that both theories
cannot be simultaneously correct. If this should
ever turn out to be true, it would be one of the most
important discoveries in the history of physics. So
far, every experiment done has shown that both

theories are correct. But more fundamentally, Goldstein appears to believe that Bohm's theory is at odds with relativity because the "influences" travel at superluminal speeds. Bohm, and the other physicists who work on this, are insistent that nothing in this theory that carries information can move faster than light. Whether one can make a relativistic version of the theory is a subject of active investigation. The same passage says that in quantum mechanics, "events can have instantaneous influences on other, far-distant events." What is an "event" and what is an "influence"? Bell was fond of the following little example. On the death of Queen Elizabeth, Prince Charles instantaneously becomes king. Surely this is an "event" having an "instantaneous influence" on another, perhaps far-off event. No one would argue that there is anything especially odd about this, or that it violates the theory of relativity. If Prince Charles were to learn instantaneously of the death of his mother, that would be another matter. I think one could make the case that the quantum mechanical situation is something like this. Probabilities change instantaneously, but we cannot communicate this with signals that travel faster than the speed of light.

Let us return to the Stern-Gerlach experiment.

So long as the singlet state of the two spinning electrons is not interfered with, the spins remain entangled. This is reflected in the correlation of the two spins. No matter what the orientation of the magnets is, each observer will observe a random pattern of spin ups and spin downs. Until the observers compare notes, they have no evidence of the correlation. Once they compare notes they see that the correlations changed when one of the magnets was rotated, even when the electrons were in flight. There is no contradiction with relativity or anything else except common sense. Given our experience with most other things, it is natural to demand an "explanation" for this bizarre behavior. It is in these explanations that the troubles begin— that one begins to introduce the "instantaneous propagation of influences." If we are willing simply to accept the quantum mechanical prediction without seeking an "explanation," then these issues do not arise.

Bohmian-Mallachian mechanics offers an explanation, but at a cost. There are instantaneous influences. But do these "influences" violate the theory of relativity? On this matter, Bohm was categorical. In the second of his two papers, he wrote: "The reason why no contradictions with relativity

arise in our interpretation despite the instantaneous transmission of momentum between particles is that no signal can be carried this way."[11] He
does not tell us what a "signal" is, nor does he tell
us whether there is a problem with causality. Of
course we now know, thanks to Bell, that these
hidden-variable theories, if they are to produce results in agreement with quantum mechanics, must
be "nonlocal." In his argument for his inequality,
Bell makes it very clear what he means by nonlocal. There is a hidden variable governing what
happens at each magnet, but the values at one
magnet do not depend in any way on the other.
This work by Bell makes whatever task the three
characters in Goldstein's novel are pursuing much
more difficult. In fact, I cannot tell from the novel
what they are trying to do. But I am pleased to report that, whatever it was, they succeed and that
Dana submits a paper with Childs as the first author. As the novel ends, she is awaiting the referee
reports.

Houellebecq is quite something else. It is not possible to imagine Goldstein's novel without the quantum theory, but it is perfectly possible to imagine
Houellebecq's novels without it. Houellebecq intro-

duces the quantum theory because it interests him. Where does this scientific interest come from? Michel Houellebecq was born Michel Thomas on the island of La Réunion on the twenty-sixth of February 1956. His mother, Janine Ceccaldi Thomas, was a medical doctor and something of a free spirit. She married René Thomas in 1953, but the marriage broke up when she had an affair with another man and gave birth to a daughter. Michel was sent to live with his maternal grandparents, but in 1961 his father reinstalled him with Michel's paternal grandmother, whose maiden name was Houellebecq. Michel Thomas adopted this name when he began his literary career. He also altered his birthdate to 1958.

Young Michel was a precocious child—always first in his class. He was a natural to enter one of the Grandes Ecoles. He chose to orient himself toward biology, which did not exempt him from two years of preparation for the entrance examinations. This preparation, which was intense, involved physics and, above all, mathematics. In the event, he chose to enter one of the less prestigious of the Grandes Ecoles: the Institut National Agronomique, which trained people to work in agriculture. As a specialty, Michel chose ecology. He graduated in 1978, and for the next three years, despite submitting his résumé innumerable times, he could not find a job. So he entered a film school, where he remained for two years. He then began a career as a computer systems analyst, eventually working at the French National Assembly. Meanwhile, he was writing poetry and a first novel (really a novella), entitled *Extension du domaine de la lutte*, which deals with a computer scientist. It has hints of the graphic sex that figures in all his novels, as well as evidence of a mordant view of human nature. It was based in part on his fellow workers, who had no idea, until it was pointed out to them, that the author Michel Houellebecq was their former colleague Michel Thomas. It was his second novel, *Les Particules élémentaires*, published in 1998, that convinced the French literary establishment that they had a supernova on their hands.

Les Particules élémentaires is about two stepbrothers, Michel Djerzinski and Bruno Ceccaldi. Houellebecq gives their mother the name Janine Ceccaldi, the name of his own estranged mother. Michel Djerzinski begins his scientific career as a physicist. Houellebecq could have made him any kind of a physicist, since he soon leaves physics

for biology, but the choice is very interesting. Djerzinski is the youngest member of Alain Aspect's
group at Orsay, which is in the process of testing
Bell's inequalities. Incidentally, Aspect told me that
Houellebecq never consulted him. But Houellebecq
gives an excellent account of the significance of the
experiments.

Aspect's experiments—precise, rigorous, and
perfectly documented—were to have profound re-
percussions in the scientific community. The results, it was acknowledged, were the first clear-cut
refutation of Einstein, Podolsky, and Rosen's objections when they claimed in 1935 that "quantum theory is incomplete." Here was a clear viola-
tion of Bell's inequalities—derived from Einstein's
hypotheses—since the results tallied perfectly with
quantum predictions. This meant that only two
hypotheses were possible. Either the hidden properties which governed the behavior of subatomic
particles were nonlocal—meaning they could instantaneously influence one another at an arbitrary distance—or the very notion of particles
having intrinsic properties in the absence of observation had to be abandoned. The latter opened up
a deep ontological void—unless one added a radical positivism and contented oneself with develop-
ing a mathematical formalism which predicted the
observable and gave up on any idea of underlying

reality. Naturally, it was the last option which won over the majority of researchers.[12]

From my own experience, I would say that the "majority of researchers" prefer not to think about the problem at all.

It is clear from Houellebecq's latest novel, *La Possibilité d'une île*, published in 2005, that his interest in the quantum theory has persisted. In this novel, which is a kind of science fiction, quantum theory is even more irrelevant. Houellebecq simply inserts his reflections on the matter when it suits him. What has caught his fancy is a development by Murray Gell-Mann and his former student James Hartle. Hartle is a frequent collaborator of Stephen Hawking. Gell-Mann won the 1969 Nobel Prize in Physics for his work on elementary particles. In 1994, Gell-Mann published a popular book on quantum theory, *The Quark and the Jaguar.* It is clear from the novel that Houellebecq has read this book. Gell-Mann and Hartle, who are old friends of mine, told me they were surprised to find their work discussed in this novel and had never had any contact with Houellebecq.

The development that led to the work of Gell-Mann and Hartle began with a paper by Dirac which he published in 1933. It was pretty much ignored until Feynman took up Dirac's ideas for a thesis that he wrote under the supervision of John Wheeler. It was finished in 1942. Feynman then went off to Los Alamos, so he did not publish anything about it until 1946. The essential idea is this. Suppose at some time t, within the limits of the uncertainty principle, measurements have been performed on such things as the position and momentum, and perhaps the spin, of some particle. The particle then evolves in time while undergoing interactions with various forces. At a later time, these same quantities are measured. If this was classical physics we could predict with certainty the results of these measurements. But in quantum mechanics all we can do is to predict probable values. In fact, in the usual interpretation of the theory, none of these properties have any value until a measurement is made. In classical physics, such a particle would trace out an orbit and we could be certain that there was such an orbit even though we do not chose to observe it at every point. In the usual interpretation of the quantum theory, we have no right to assume anything about orbits in the absence of measurement. This means that the particle could have had several possible "paths" or "histories" between the times that we made our observations. In computing the probabilities, we have to sum over all these histories. This being quantum mechanics, these histories can interfere with each other the way that waves do. But interactions can cause these histories to "decohere." Indeed, under the right circumstances these decoherent histories behave at least approximately like classical orbits. This is how, according to this picture, the classical world emerges from the quantum theory.

Some adaptive systems—such as us—are capable of observing. Gell-Mann and Hartle call such systems "IGUS": Information-Gathering and -Utilizing Systems. As such a system evolves, it confronts different possible branch histories but follows only one. The others it discards. In *The Quark and the Jaguar*, Gell-Mann writes: "An observation in this context means a kind of pruning of the tree of branching histories. At a particular branching, only one of the branches is preserved (more precisely, on each branch only that branch is preserved!). The branches that are pruned are thrown away, along with all the parts of the trees that grow out of the branches that are pruned."[13] Wave functions don't collapse. The discarded bits belong to another history.

All of this seems to have appealed to Houelle-becq's poetic imagination. He writes in his novel *La Possibilité d'une île* (the translation is mine): "For an IGUS observer, be it natural or artificial, only one branch of the universe can be given a real existence. Although this conclusion does not at all exclude the possibility of other branches of the universe, it denies the given observer access to them. In the formulation, at once mysterious and synthetic, of Gell-Mann: 'On each branch, only that branch is preserved.' The presence even of one community of observers reduced to two IGUS constitutes the proof of the existence of a reality."[14]

While Houellebecq finds poetry in quantum mechanics, others have found spirituality. Here a word of caution is in order. Schrödinger was a student of oriental religions and wrote about them. But this is what he said in the introduction to his book *My View of the World:*

There is one complaint which I shall not escape. Not a word is said here of acausality, wave mechanics, indeterminacy relations, complementarity, an expanding universe, continuous creation, etc. Why doesn't he talk about what he knows instead of trespassing on the professional philosopher's preserves? *Ne sutor supra crepidam.*

[Cobbler stick to your last.] On this I can cheerfully justify myself: because I do not think that these things have as much connection as is currently supposed with a philosophical view of the world.[15]

I am going to conclude this chapter, and the book, by discussing people who feel that "these things" have everything to do with a philosophical view of the world.

So far as I can tell, the present vogue began when Bell's work became widely known. For example, in 1975 the physicist Henry Stapp wrote that "Bell's theorem is the most profound discovery of science."[16] One might have thought that Stapp would have put the discovery of quantum mechanics itself somewhat higher on the list. But Stapp's evaluation is widely quoted by people who use Bell's work as an argument for various mystical beliefs. Among the first was Gary Zukav, who in 1979 published *The Dancing Wu Li Masters*.[17] In the opening pages, Zukav, who is not a physicist, informs us that he got his inspiration for the book when he attended a conference at the Esalen Institute at Big Sur in California. Esalen is better known for teaching New Age mysticism than physics—but if the two can be mixed, the sky is the limit. Inspired by his visit, Zukav began talking to several physicists, including the aforementioned Henry Stapp. The result of all of this is a popular presentation of such things as elementary-particle physics as they were in the mid-1970s. The problem is that thirty years have gone by, and in this field that is a millennium. A new generation of physicists is produced roughly every ten years. This means that physics is now being done by the graduate students of the graduate students of the physicists Zukav interviewed. Even though I have the 2001 edition of his book, the most up-to-date reference is 1977! Reading this book is like being in a time warp. Indeed, that is the risk of trying to link some set of spiritual beliefs to a contemporary science. It is like building a house on a block of ice. The Dalai Lama, whose interest in science is very contemporary, has said on many occasions—I heard him say it myself—that if there was a conflict between a Buddhist tenet and a proven scientific fact, the Buddhist tenet would have to be modified. What is one to make of a statement like following (it is typical) in Zukav:

> Subatomic particles forever partake of this unceasing dance of annihilation and creation. In fact, subatomic particles *are* this unceasing dance of

annihilation and creation. This twentieth-century discovery, with all its psychedelic implications, is not a new concept. In fact, it is very similar to the way that much of the earth's populations, including the Hindus and the Buddhists, view their reality.

Hindu mythology is virtually a large-scale projection into the psychological realm of microscopic scientific discoveries. Hindu deities such as Shiva and Vishnu continually dance the creation and destruction of universes, while the Buddhist image of the wheel of life symbolizes the unending process of birth, death, and rebirth which is part of the world of form, which is emptiness, which is form.[18]

I have a test for phrases like "which is part of the world of form, which is emptiness, which is form." I negate the propositions: "which is not part of the world of form, which is not emptiness, which is not form." If I cannot attach more sense to one as opposed to the other, I go on to something else. Try as I might, I cannot make much sense out of Zukav's book. I also cannot make much sense out of Fritjof Capra's book *The Tao of Physics*, published in 1975.

Capra has a Ph.D. in physics. The theory of elementary particles he worked on thirty or more years ago had, at the time, a great vogue. Now no one has much interest in it, although you would not be able to tell this from reading his book. On the other hand, he has to a certain extent kept up with developments, and the latest edition of his book (1999) reflects this. In an afterward to the most recent edition, Capra says he is pleased that no physicist has found anything wrong in his presentation. Pauli had a category of things that were not even wrong—they were not coherent enough for anyone to assign truth or falsity to them. Here is a fairly typical paragraph in Capra:

> The conception of physical things and phenomena as transient manifestations of an underlying fundamental entity is not only a basic element of quantum field theory, but also a basic element of the Eastern world view. Like Einstein, the Eastern mystics consider this underlying reality as the only reality; all its phenomenal manifestations are seen as transitory and illusory. This reality of the Eastern mystic cannot be identified with the quantum field of the physicist, because it is seen as the essence of all phenomena in this world, and consequently is beyond all concepts and ideas. The quantum field, on the other hand, is a well-defined concept which only accounts for some of the physical phenomena. Nevertheless, the intuition behind the physicist's interpretation of the subatomic

world, in terms of the quantum field, is closely paralleled by that of the Eastern mystic who interprets his or her experience of the world in terms of an underlying reality. Subsequent to the emergence of the field concept, physicists have attempted to unify the various fields into a single fundamental field which would incorporate all physical phenomena. Einstein, in particular, spent the last years of his life searching for such a unified field. The *Brahman* of the Hindus, like the *Dharmakaya* of the Buddhists, and the *Tao* of the Taoists, can be seen, perhaps, as the ultimate unified field from which spring not only the phenomena studied in physics, but all other phenomena as well.[19]

I especially like the "perhaps" in the last sentence. Capra's book has had many readers who admire it. I am not one of them.

The ne plus ultra of this vogue is the film *What the #$*! Do We Know?* It is a docudrama that combines interviews, animation, and scenes using actors. When I first saw the film, and especially the interviews, I had a sense of déjà vu. In the 1960s I spent a good deal of time in a coffeehouse-restaurant in Greenwich Village called the Café Figaro. The food was good, the prices were reasonable, and the owners allowed you to play chess.

You could go to any table in the Figaro and find people who talked and looked like the ones in the film. I wonder if some of them were the same people. At any table—including, I am sure, my own—the most amiable nonsense was being propounded with the greatest sincerity, although I do not remember the word "quantum" ever being used by anybody. Most of us grew out of it, but not the people in this film. Here is another sampling to go with the ones at the head of the chapter.

> To acknowledge the quantum self is to acknowledge the place where you rarely have chosen to acknowledge mind. . . . When that shift of perspective takes place, we say that somebody has been enlightened.
>
> —Amit Goswami, Ph.D.[20]

And finally:

> We're living in a world where all we see is the tip of the iceberg. The immense tip of the quantum mechanical iceberg.
>
> —John Hagelin, Ph.D.[21]

Well, maybe so.

L'Envoi

WHEN I STARTED this exploration, I had no idea where it would lead—that we would go from Auden to the Dalai Lama and on to Communist ideology. The riches of the quantum theory are so vast that I am sure we are closer to the beginning than to the end. But here are two of my favorite quotations, one from Bohr and the other from Feynman. They are a good note to end on. Bohr wrote:

> Those who are not shocked when they first come across quantum theory cannot possibly have understood it.

And Feynman:

> Might I say immediately . . . we always have had a great deal of difficulty in understanding the world view that quantum mechanics represents. . . . I cannot define the real problem, therefore I suspect there's not a real problem, but I am not sure there's no real problem.

Notes

Index

Notes

1. Bishops

1. Quoted in John Barnes, *Ahead of His Age: Bishop Barnes of Birmingham* (London: Collins, 1979), 190.
2. Ibid., 344.
3. Ibid., 194.
4. I am grateful to Edward Mendelson for this information and for other insights into the poem.
5. Ernest William Barnes, *Scientific Theory and Religion: The World Described by Science and Its Spiritual Interpretation* (Cambridge: Cambridge University Press, 1933).
6. Ibid., 458.
7. Ibid.
8. Ibid., 409.
9. Ibid., 191–192.

2. Quantum Buddhists

1. Dalai Lama, *The Universe in a Single Atom: The Convergence of Science and Spirituality* (New York: Morgan Road Books, 2005), 2–3.
2. Ibid., 55–56.

3. I am grateful to Luis Alvarez-Gaumé and Pierre Jarron of CERN for providing me with a DVD of this film.

4. In the film, Bell actually reads a quote from Joseph Needham to this effect.

5. Dalai Lama, *The Universe in a Single Atom*, 34–35.

6. David Bohm, *Quantum Theory* (New York: Prentice-Hall, 1951).

7. I am grateful to Bruce Rosenblum for telling me that, on a visit to Princeton, he heard Einstein say that "David" had not done what he had asked him to do. Einstein did not explain what he had asked Bohm to do.

8. Dalai Lama, *The Universe in a Single Atom*, 30.

9. Ibid., 128.

10. This statement by the Dalai Lama does not make it clear that these particles have to be created in what is called an "entangled" state. This is explained in Chapter 6.

11. Dalai Lama, The Universe in a Single Atom, 64–66.

3. Léon Rosenfeld

1. This and much of the biographical material on Rosenfeld I have taken from Anja Skaar Jacobsen, "Léon Rosenfeld's Marxist Defense of Complementarity," *Historical Studies in the Physical and Biological Sciences*, 37, Supplement (2007): 3–37, quotation on 4. There does not seem to be a full biography of Rosenfeld. Pauli's joke was in part mathematical.

He was saying that Rosenfeld was the geometric mean between Bohr and Trotsky.

2. Ibid., 24.

3. Abraham Pais, *Niels Bohr's Times* (New York: Oxford University Press, 1991), 9–10.

4. Ibid.

5. V. I. Lenin, *Materialism and Empirio-criticism* (Moscow: Progress Publishers, 1987; orig. pub. 1909), 79.

6. Ibid., 148.

7. Pais, *Niels Bohr's Times*, 430–431.

8. I am grateful to Anja Skaar Jacobsen for this information.

9. Robert Oppenheimer, *Letters and Reflections*, ed. Alice Kimball Smith and Charles Wiener (Cambridge, Mass.: Harvard University Press, 1980), 190.

10. Louis de Broglie, "Note," *Comptes rendus de l'Académie des Sciences*, 177 (1923): 507–510.

11. The first and third are "The transformation of quantity into quality" and "The negation of the negation."

12. For a very enlightening discussion of these matters, see Loren R. Graham, "Quantum Mechanics and Dialectical Materialism," *Slavic Review*, 3 (September 1966): 381–410; and the response by Vladimir Fock, ibid., 411–413.

13. Graham, "Quantum Mechanics," 384.

14. Ibid., 392.

15. Fock, Response to Graham, 412.

16. Henry Stapp, "The Copenhagen Interpretation," *American Journal of Physics*, 40, no. 8 (1972): 1098–1116.

17. Ibid., 1115.

18. Jacobsen, "Léon Rosenfeld's Marxist Defense," 25.

19. For a very interesting account of Bohm in Brazil, see Olival Freire Jr., "Science and Exile: David Bohm, the Hot Times of the Cold War, and His Struggle for a New Interpretation of Quantum Mechanics," *Historical Studies on the Physical and Biological Sciences*, 36, no. 1 (2005): 1–34. See also F. David Pate's biography of Bohm, *Infinite Potential* (New York: Basic Books, 1997). The reader should be aware that this book contains many mistakes of detail. For example, Pate claims that Bohm and Richard Feynman were both graduate students of Oppenheimer's. Feynman was in fact a student of John Wheeler at Princeton. Allen Shenstone has become Allen Shelstone, Stirling A. Colgate is referred to as Stephen A. Colgate, etc., etc.

20. I am grateful to David Pines (who was one of those students) for this information.

21. Freire, "Science and Exile," 20.

22. Jacobsen, "Léon Rosenfeld's Marxist Defense," 26.

4. A Double Slit

1. Richard P. Feynman, *The Feynman Lectures on Physics*, ed. Robert B. Leighton and Matthew Sands,

3 vols. (Reading, Mass.: Addison-Wesley, 1963–1965), vol. 3, ch. 37.

2. Tom Stoppard, *Hapgood*, in Stoppard, *Plays* (London: Faber and Faber, 1999), 544.

3. I am grateful to the Newton scholar Alan Shapiro of the University of Minnesota for clarifying remarks on Newton, Huygens, and Young.

4. Albert Einstein, *The Meaning of Relativity* (Princeton, N.J.: Princeton University Press, 1950), 1.

5. See, for example, en.wikipedia.org/wiki/Double-slit_experiment for a review with many additional references.

6. A reader who wants to know the details of this improbable saga can consult my autobiographical memoir *The Life It Brings* (New York: Penguin, 1988).

7. Tom Stoppard, "The Matter of Metaphor," *New Theater Review*, 11 (Fall 1994): 4.

8. Ibid.

9. Technically, the wave function is a so-called complex number, so you must take the square of what is known as the absolute value.

10. Stoppard, *Hapgood*, 500–501.

5. A Measurement

1. John von Neumann, *Mathematical Foundations of Quantum Mechanics* (Princeton, N.J.: Princeton University Press, 1955), 419–420.

2. Amit Goswami, *The Self-Aware Universe* (New York: Penguin Putnam, 1993), 205.

3. Niels Bohr, *The Philosophical Writings of Niels Bohr*, vol. 4 (Woodbridge, Conn.: Ox Bow Press, 1998), 81.

4. David Bohm, "A Suggested Interpretation of the Quantum Theory in Terms of 'Hidden Variables,'" *Physical Review*, 85 (1952): 166–193. Reprinted in J. A. Wheeler and W. H. Zurek, eds., *Quantum Theory and Measurement* (Princeton, N.J.: Princeton University Press, 1983).

5. Hugh Everett, "'Relative State' Formulation of Quantum Mechanics," *Reviews of Modern Physics*, 29 (1957): 454–462.

6. Eugene Wigner, "Remarks on the Mind-Body Question" (1961), in J. A. Wheeler and W. H. Zurek, eds., *Quantum Theory and Measurement* (Princeton, N.J.: Princeton University Press, 1983).

7. Ibid., 169.

8. Ibid., 172–173.

9. Ibid., 173.

6. Entanglements

1. Erwin Schrödinger, "The Present Situation in Quantum Mechanics" (1935), trans. John D. Trimmer, in J. A. Wheeler and W. H. Zurek, eds., *Quantum Theory and Measurement* (Princeton, N.J.: Princeton University Press, 1983), 161.

2. Gary Zukav, *The Dancing Wu Li Masters* (New York: Perennial Classics, 2001; orig. pub. 1979), 335.

3. Fritjof Capra, *The Tao of Physics* (Boston: Shambhala, 2000; orig. pub. 1975), 313.

4. Albert Einstein, "Über die vom Relativitätsprinzip geforderte Trägheit der Energie" (On the Inertia of Energy Required by the Relativity Principle), *Annalen der Physik*, 23, no. 7 (1907): 371–384.

5. This term was introduced by the late Gerald Feinberg, who was a professor at Columbia University. He actually participated in an unsuccessful attempt to find them. ·

6. John Bell, *Speakable and Unspeakable in Quantum Mechanics* (New York: Cambridge University Press, 2004), 235–236.

7. Max Born, *Natural Philosophy of Cause and Chance* (New York: Dover, 1964), 109.

8. John Bell, "On the Problem of Hidden Variables in Quantum Mechanics," *Reviews of Modern Physics*, 38 (1966): 447–452.

9. Ibid., 447, 350.

10. For small angles, $\cos(a) \sim 1 - a^2/2$.

11. *The Born-Einstein Letters* (New York: Walker, 1971), 192.

12. John Bell, "On the Einstein Podolsky Rosen Paradox," *Physics*, 1 (1965): 195–200.

13. J. F. Clauser, M. A. Horne, A. Shimony, and R. A. Holt, "Proposed Experiment to Test Local Hidden-Variable Theories," *Physical Review Letters*, 23 (1969): 880–884.

14. encyclopedia.kids.net.au/page/jo/John_Stewart_Bell.

15. beliefnet.com/story/136/story_13663_2.html.

7. Anyway, What the #$*! Do We Know?

1. Stuart Hameroff, in William Arntz and Betsy Chasse, eds., *The Little Book of Bleeps*, quotations from the movie *What the #$*! Do We Know?* (Captured Light Distribution, 2004). This book, perhaps inspired by the quantum theory uncertainty principle, does not offer page numbers.

2. Jeffrey Satinover, quoted in *The Little Book of Bleeps*.

3. Rebecca Goldstein, *Properties of Light* (Boston: Houghton Mifflin, 2001), 37–38.

4. Michel Houellebecq, *The Elementary Particles*, trans. Frank Wynne (New York: Knopf, 2000), 103. In French, *Les Particules élémentaires* (Paris: Flammarion, 1998).

5. "Ein Abend mit Lawrence Durrell," interview with Lawrence Durrell by Dieter Zimmer, *Die Zeit*, 48 (November 27, 1959). In English as "Becoming a Literary Tramp," in *Lawrence Durrell: Conversations*, ed. Earl G. Ingersoll (Cranbury, N.J.: Associated University Press, 1998), 37–38.

6. Goldstein, *Properties of Light*, 41–42.

7. John Bell, *Speakable and Unspeakable in Quantum Mechanics* (New York: Cambridge University Press, 2004), 160.

8. Goldstein, *Properties of Light*, 49.

9. Ibid., 53.

10. Ibid., 71–72.

11. David Bohm, "A Suggested Interpretation of the

Quantum Theory in Terms of 'Hidden Variables,'" *Physical Review*, 85 (1952); reprinted in J. A. Wheeler and W. H. Zurek, eds., *Quantum Theory and Measurement* (Princeton, N.J.: Princeton University Press, 1983), 390.

12. Houellebecq, *The Elementary Particles*, 103.

13. Murray Gell-Mann, *The Quark and the Jaguar* (New York: W. H. Freeman, 1994), 155.

14. Michel Houellebecq, *La Possibilité d'une île* (Paris: Fayard, 2005), 345.

15. Erwin Schrödinger, *My View of the World*, trans. Cecily Hastings (Cambridge: Cambridge University Press, 1964), vii–viii.

16. Henry P. Stapp, "Bell's Theorem and World Process," *Il nuovo cimento*, 29B, no. 2 (1975): 271.

17. Gary Zukav, *The Dancing Wu Li Masters* (New York: Perennial Classics, 2001; orig. pub. 1979).

18. Ibid., 240–241.

19. Fritjof Capra, *The Tao of Physics* (Boston: Shambhala, 2000; orig. pub. 1975), 211.

20. Amit Goswami, quoted in *The Little Book of Bleeps*.

21. John Hagelin, quoted ibid.

Index